7

Nutrient Deficiencies & Toxicities In Crop Plants

edited by

William F. Bennett

College of Agricultural Sciences and Natural Resources
Texas Tech University, Lubbock

APS PRESS

The American Phytopathological Society
St. Paul, Minnesota

**Financial support for this book
was provided by**

Farmland Industries, Inc.
Foundation for Agronomic Research
IMC Fertilizer, Inc.
Kerley Enterprises, Inc.

Reference in this publication to a trademark, proprietary product,
or company name by personnel of the U.S. Department of Agriculture
or anyone else is intended for explicit description only and does not
imply approval or recommendation to the exclusion of others that
may be suitable.

Library of Congress Catalog Card Number 93-71744
International Standard Book Number 0-89054-151-5

Printed in the United States of America on acid-free paper

The American Phytopathological Society
3340 Pilot Knob Road
St. Paul, Minnesota 55121-2097, USA

Preface

The purpose of this book is to present the symptoms associated with essential nutrient deficiencies and excesses (toxicities). These symptoms can be used as one means of determining deficiencies or toxicities and, therefore, fertilizer and/or soil amendment need. Symptoms, however, need to be used with other diagnostic tools including soil tests and plant tissue analyses plus information on the farmer's management practices.

Plant growth abnormalities are an indicator or symptom of one or more unfavorable growing conditions. The condition may be a lack or an excess of an essential plant nutrient or an excess of a nonessential element. It may be an unsatisfactory environment—too cold or hot, too wet or dry, the lack of oxygen or CO_2, too much SO_2 in the atmosphere, or too much wind. It may be soil characteristics—poor physical conditions, improper soil pH, or saline or sodic soil. It may be too much of a herbicide. It could be a biological hazard such as disease, insects, or even birds and larger animals. Man-made activities can cause problems such as root pruning during cultivation. Many of the abnormal growth patterns, or symptoms, may be specific enough to pinpoint a nutrient deficiency or a toxicity. Oftentimes, however, it is difficult to ascertain the exact cause of the abnormality.

The premier and excellent publication, *Hunger Signs in Crops* edited by H. B. Sprague, has been the most-used source of information on nutrient deficiency symptoms. With that book now out of print, it seemed appropriate to prepare a new guide to nutrient deficiency symptoms and to include a compilation of toxicity symptoms. Toxicities due to an excess of an essential nutrient or a nonessential element have been known for many years. There is much more interest now in this topic, since problems due to heavy metals are becoming more prevalent.

My own interest in wanting to prepare this book grew out of teaching the undergraduate and graduate Soil Fertility courses for 25 years at Texas Tech University, using as a text the very fine book entitled *Soil Fertility and Fertilizer Use* initially written by my personal friend, Sam Tisdale, and other coauthors. Other experiences which prompted me to want to prepare this book included my years of work as Chief Agronomist with Elcor Chemical/Western Ammonia fertilizer manufac-

turer, plus my role as State Extension Soil Chemist with the Texas Agricultural Extension Service.

An excellent feature of this book is the more than 300 colored plates that depict the symptoms of nutrient deficiencies and toxicities. The book is also unique because of the large number of toxicity symptoms depicted.

This book should be of value to the ever-increasing force of agronomic consultants. Producers can use it in diagnosing nutrient problems. Fertilizer dealers, agrichemical dealers and other suppliers of inputs to producers could benefit. It should be of value to educators—teachers of high school agricultural science classes; county extension advisors; Soil Conservation Service agronomists; and university scientists.

Many individuals need to be acknowledged:

The authors (and coauthors) of the chapters on 19 different crops.

The many co-workers of the authors who furnished slides depicting plant symptoms.

Larry Sanders, Regional Director of the Potash-Phosphate Institute, Stanley, Kansas, who reviewed in a most capable and thorough manner all of the chapters for subject matter content.

Arthur McCain, University of California, Berkeley, who acted as liaison for the APS Press on this publication.

Administrative Assistant Claudia Thornton for her capable and diligent assistance in coordinating the handling and transmitting of the book materials.

Those who provided secretarial assistance during the preparation of manuscripts and editing process.

Many others were involved in important ways, and to all of these go a big thank you.

In my career in teaching, research, and extension work, I have always wanted to acknowledge those many scientists and researchers down through the years who have developed the body of knowledge necessary to prepare this type of book. Much of the information is generally known, and it is virtually impossible to give credit to all who deserve it.

William F. Bennett

Plate Credits

Copyright is not claimed for photographs reprinted from the APS Disease Compendium Series that were created by U.S. government employees as a part of their official duties.

Chapter 2, Corn. **1 and 2** courtesy R. A. Wiese; **3, 4, 6, and 13** courtesy Regis D. Voss; **5 and 8–10** courtesy D. A. Whitney; **7, 15, and 16** courtesy R. G. Hoeft; **11** courtesy D. D. Buchholz; **12** courtesy L. Bundy; **14** courtesy L. F. Welch; **17 and 18** courtesy C. Harms

Chapter 3, Rice. **19–21, 23, 25, 26, and 29–31** courtesy authors of Chapter 3; **22** copyright Potash and Phosphate Institute (used by permission); **24, 27, and 28** courtesy George Snyder

Chapter 4, Sorghum. **32–34, 37–39, 41–47, 49–52, and 54–63** courtesy R. B. Clark; **35, 36, 40, 48, and 53** reprinted from Compendium of Sorghum Diseases, edited by Richard A. Frederiksen, American Phytopathological Society, St. Paul, MN, 1986

Chapter 5, Wheat and Other Small Grains. **64** reprinted from Compendium of Wheat Diseases, 2nd ed., by M. V. Wiese, American Phytopathological Society, St. Paul, MN, 1987; **65, 66, 68, 73, and 74** courtesy R. Mahler; **67 and 69** reprinted, by permission, from Nutrient Deficiencies and Toxicities in Wheat: A Guide for Field Identification, by K. Snowball and A. D. Robson, CIMMYT, Mexico City, 1991; **70** courtesy T. G. Atkinson; **71 and 72** courtesy G. Murray; **75** courtesy M. V. Wiese

Chapter 6, Sugarcane. **76–89** courtesy authors of Chapter 6

Chapter 7, Sugar Beet. **90, 91, 93, 102, 105, 111, 114, 116, 117, 120, and 123** courtesy authors of Chapter 7; **92, 94–101, 103, 104, 106–110, 112, 113, 115, 118, 119, 121, 122, and 124–128** reprinted from Compendium of Beet Diseases and Insects, edited by E. D. Whitney and James E. Duffus, American Phytopathological Society, St. Paul, MN, 1986

Chapter 8, Soybeans. **129–138** reprinted from Compendium of Soybean Diseases, 3rd ed., edited by J. B. Sinclair and P. A. Backman, American Phytopathological Society, St. Paul, MN, 1989

Chapter 9, Peanuts. **139 and 143** courtesy J. Beasley; **140** courtesy D. Gorbet; **141** courtesy P. Gillier; **142** courtesy D. Hartzog; **144, 146, and 147** reprinted from Compendium of Peanut Diseases, edited by D. Morris Porter, Donald H. Smith, and R. Rodríguez-Kábana, American Phytopathological Society, St. Paul, MN, 1984; **145 and 148** courtesy D. Smith; **149** courtesy F. Boswell

Chapter 10, Cotton. **150** courtesy D. M. Bassett; **151 and 152** courtesy K. G. Cassman; **153, 155, 157, 159, and 160** copyright Potash and Phosphate Institute (used by permission); **154** courtesy A. E. Ludwick; **156** courtesy T. A. Kerby; **158** courtesy B. R. Wells

Chapter 11, Cucurbits: Cucumber, Muskmelon, and Watermelon. **161–164** courtesy Salvadore J. Locascio

Chapter 12, Onions. **165, 169, and 172** courtesy L. A. Ellerbrock; **166–168, 170, and 171** courtesy D. D. Warncke

Chapter 13, Tomato. **173–190** courtesy Gerald E. Wilcox

Chapter 14, Common Bean. **191 and 193–196** reprinted from Compendium of Bean Diseases, edited by Robert Hall, American Phytopathological Society, St. Paul, MN, 1991; **192** courtesy H. F. Schwartz

Chapter 15, Potato. **197–200, 202–212, and 214–232** courtesy Albert Ulrich; **201 and 213** reprinted from Compendium of Potato Diseases, edited by W. J. Hooker, American Phytopathological Society, St. Paul, MN, 1981

Chapter 16, Apples and Pears. **233 and 234** courtesy Eric Hanson; **235, 242, 245, 246, and 248** courtesy D. Burkhart; **236** courtesy A. Ludwick; **237** courtesy D. Dewey; **238, 239, and 244** courtesy W. Stiles; **240, 243, 247, and 249** courtesy F. Peryea; **241** courtesy F. Dennis

Chapter 17, Citrus. **250, 251, 253, 254, 256–265, 267, and 268** courtesy authors of Chapter 17; **252** courtesy T. W. Embelton; **255** courtesy R. C. J. Koo; **266** courtesy W. Reuther; **269** reprinted from Compendium of Citrus Diseases, edited by J. O. Whiteside, S. M. Garnsey, and L. W. Timmer, American Phytopathological Society, St. Paul, MN, 1988

Chapter 18, Stone Fruit: Peaches and Nectarines. **270, 271, 275, and 277** courtesy F. T. Yoshikawa; **272 and 276** courtesy J. H. Larue; **273, 278, and 281** courtesy R. Scott Johnson; **274 and 279** courtesy N. F. Childers; **280** courtesy R. L. Stebbins

Chapter 19, Grapes. **282–299** courtesy Wilhelm Gärtel

Chapter 20, Turfgrass. **300–303** courtesy T. R. Turner; **304** courtesy D. V. Waddington; **305–307** courtesy D. D. Minner

Contents

Chapter 1

William F. Bennett
College of Agricultural Sciences and Natural Resources
Texas Tech University, Lubbock

Plant Nutrient Utilization and Diagnostic Plant Symptoms

Essential Elements

Early-day scientists discovered that chemical elements are utilized by plants and are essential for plant growth. At one time, water was thought to be the sole nutrient needed by plants. It was later found that certain "earth substances" were beneficial for plant growth. In the 1800s, Von Liebig supported the essential role of certain elements and proposed that growth is limited to the extent that an essential element is lacking (20).

As early as the 1930s, 14 elements were known to be essential (13,16). Carbon (C), hydrogen (H), and oxygen (O), the "building block" elements, were known for many years before that time to be required for photosynthesis to occur. Six other elements known to be essential, called primary and secondary elements at that time, were nitrogen (N), phosphorus (P), potassium (K), calcium (Ca), magnesium (Mg), and sulfur (S). Other elements, called trace elements, included iron (Fe), zinc (Zn), manganese (Mn), copper (Cu), and boron (B) (13).

In the 1950s, chlorine (Cl) and molybdenum (Mo) were added to the list (6,7). Questions were raised about the essentiality of silicon (Si), sodium (Na), vanadium (V), and cobalt (Co). These four were listed as essential by 1966 (8,20). The terms *macronutrients*, those elements needed in relatively large quantities, and *micronutrients*, essential elements needed in relatively small amounts, began to replace the words primary, secondary, and trace or minor elements (7). Tisdale et al (20) now lists 20 elements that are essential for plant growth.

Criteria for the essentiality of elements for plant growth are not always consistent. Arnon, in 1939, was one of the first to establish criteria for essentiality of nutrients. Meyer and Anderson listed criteria (16) that agree closely with those of Arnon. They pointed out that essentiality is demonstrated only if lack of it can be shown to result in injury, abnormal development, or death of plants grown in sand or solution culture. According to Nicholas, if an element is a "functional or metabolic nutrient," it is essential (20).

From a practitioner's viewpoint, essentiality of any particular element may not be too important. Of more concern is whether the element is of economic importance to the producer.

Except in highly intensified production in environmentally controlled structures where carbon dioxide might be limiting, the elements C, H, and O are not considered part of a producer's fertilization program. All macronutrients are of potential economic concern along with the micronutrients Fe, Zn, Mn, Cu, and B. Chlorine and Mo are now gaining prominence as important micronutrients for production of certain crops. This

Table 1.1. Nutrients essential to plant growth

Element	Chemical symbol	Form(s) taken up by plant
Macronutrients		
Carbon	C	CO_2
Hydrogen	H	H_2O
Oxygen	O	H_2O, O_2
Nitrogen	N	NH_4^+, NO_3^-
Phosphorus	P	$H_2PO_4^-$, HPO_4^{2-}
Potassium	K	K^+
Calcium	Ca	Ca^{2+}
Magnesium	Mg	Mg^{2+}
Sulfur	S	SO_4^{2-}
Micronutrients		
Iron	Fe	Fe^{2+}, Fe^{3+}
Zinc	Zn	Zn^{2+}, $Zn(OH)_2^0$
Manganese	Mn	Mn^{2+}
Copper	Cu	Cu^{2+}
Boron	B	$B(OH)_3^0$
Molybdenum	Mo	MoO_4^{2-}
Chlorine	Cl	Cl^-
Silicon	Si	$Si(OH)_4^0$
Sodium	Na	Na^+
Cobalt	Co	Co^{2+}
Vanadium	V	V^+

leaves Si, Na, Co, and V as the elements that are normally of little economic importance, except for a few major crops and in certain production areas, and which are seldom applied to plants as a fertilizer.

For this publication, the six essential macronutrients— N, P, K, Ca, Mg, and S—and five of the micronutrients— Fe, Zn, Mn, Cu, and B—will be discussed for each crop. In some crops, other elements will be discussed when they are of some economic importance.

Elements that are generally considered to be essential for plant growth are listed in Table 1.1. The chemical symbol, which will be used throughout the publication, and the form(s) taken up by the plant are also listed for each essential element.

A large number of other elements have been found in plants, and some have been shown to play a minor role in certain plants. These include the halogens— bromine (Br), iodine (I), and fluorine (F)—which occur widely in plants. Others include strontium (Sr), tungsten (W), nickel (Ni), and rubidium (Rb). Some nonessential elements, such as selenium (Se), can accumulate sufficiently to be poisonous to animals. These elements can often replace one that is essential and perform its role in certain processes (5); they are frequently called beneficial elements.

Classifications of Essential Elements

Elements that are essential for plant growth can be classified in several ways. The term *mineral* elements is often used to refer to essential elements. This is a slight misnomer in that elements, or plant nutrients, are not minerals. The term comes from the fact that most essential elements were once combined with other elements in the form of minerals, which eventually broke down into their component parts. Mineral elements would include all essential elements other than carbon, hydrogen, and oxygen, which are derived from CO_2 and H_2O, and nitrogen, which originally came from atmospheric N_2.

Another classification of the essential elements is as metals or nonmetals. Metals include K, Ca, Mg, Na, Fe, Zn, Mn, Cu, Mo, Co, and V. Nonmetals include N, P, S, B, Cl, and Si.

Elements can also be classified by general function: 1) as a constituent of either organic or inorganic compounds—N, S, P, Ca, B, Fe, and Mg; 2) as an activator, cofactor, or prosthetic group of enzyme systems—K, Mg, Ca, Fe, Zn, Mn, Cu, Mo, Na, and Cl; 3) as a charge carrier in oxidation-reduction reactions—P, S, Fe, Mn, Cu, and Mo; and 4) as an osmoregulator and for electrochemical equilibrium in cells—K, Na, and Cl.

Functions of Essential Elements

Nitrogen

Nitrogen is absorbed by plants in both the nitrate

(NO_3^-) and ammonium (NH_4^+) forms. It is generally understood that NH_4^+ is absorbed and utilized primarily by young plants, whereas NO_3^- is the principal form utilized during the grand growth period. With the use of inhibitors, plants can benefit from the NH_4^+ form, particularly late in the season.

Plants vary in the proportion of NH_4^+/NO_3^- utilization. Some plants, such as rice, utilize primarily NH_4^+, whereas most other crop plants utilize principally NO_3^-.

Nitrogen has numerous functions in the plant. The NO_3^- ion undergoes transformation after it is absorbed and is reduced to the amine form (NH_2^-). It is then utilized to form amino acids. Twenty amino acids are precursors of polypeptide chains comprising all proteins. Two other amino acids, glycine and glutamate, are precursors of nitrogen bases. Amino acids are essential for protein formation and are considered its building blocks. They are a part of the nucleic acids (DNA and RNA) that respectively hold the genetic information and direct protein synthesis (5). In addition to amino acids, proteins, nucleic acids, and N bases, nitrogen is also a constituent of other plant compounds including nucleotides, amides, and amines. Many enzymes are proteinaceous; hence, N plays a key role in many metabolic reactions.

Because nitrogen is contained in the chlorophyll molecule, a deficiency of N will result in a chlorotic condition of the plant. Nitrogen is also a structural constituent of cell walls (18).

Protein is continually being synthesized and degraded in the plant so that N moves from older plant parts to younger leaves (18); therefore, N deficiency symptoms normally appear first on older leaves. Interactions of one nutrient with others are commonplace in metabolic processes. An example is the nitrate-reduction process. Potassium controls the amount of organic acids that influence nitrate reduction. Phosphorus, through its role in enzyme formation, influences this process. Sulfur is involved through its effect on protein and enzyme formation, and molybdenum is involved in the process through the transfer of electrons (1).

Phosphorus

Phosphorus is absorbed by plants in one of two forms, either as the monovalent phosphate ion $(H_2PO_4^-)$ or as the divalent phosphate ion (HPO_4^{2-}). The ion absorbed is determined by the pH of the soil. The $H_2PO_4^-$ predominates in soil with a pH of less than 7.2 and HPO_4^{2-} at a pH greater than 7.2 (20). There is a notable effect of N on P uptake by plants. When N and P are physically and/or chemically associated in the soil, P uptake is enhanced (1).

Phosphorus is a constituent of plant compounds such as enzymes and proteins and is a structural component of phosphoproteins, phospholipids, and nucleic acids. Since it is part of nucleic acids and of genes and chromosomes, it plays a vital role in the life cycle of plants and is important in reproductive growth. It "promotes" early maturity and fruit quality. Phosphorus has been described as ubiquitous in the plant, being involved in nearly all metabolic processes.

Phosphorus is contained in nicotinamide adenine dinucleotide phosphate (NADP), a part of the photosynthetic process (17). Its best known function is in energy storage and transfer through the compounds adenosine diphosphate (ADP) and adenosine triphosphate (ATP). It is an integral part of the reproductive system as a component of the genetic memory system of RNA and DNA and is, therefore, involved in the transfer of genetic information. It is also involved in electron transport in oxidation-reduction reactions (20).

Phosphorus plays a regulatory role in the formation and translocation of substances such as sugars and starches. It is important in the maturation processes and in seed formation. It is involved in symbiotic N fixation (17).

Potassium

Potassium is required for turgor buildup in plants and maintains the osmotic potential of cells, which in guard cells governs the opening of stomata (12). This osmotic regulation indicates the role K plays in water relations in the plant. It is involved in water uptake from soil, water retention in the plant tissue, and long-distance transport of water and assimilates in the phloem and xylem (15). It also functions in pH stabilization in the cell. It counteracts the negative charge of organic acids and inorganic anions such as Cl^- and SO_4^{2-}. Potassium is required as an activator for more than 60 enzymes in meristematic tissue (19,20).

It is important in cell growth primarily through its effect on cell extension. With adequate K, cell walls are thicker and provide more tissue stability. This effect on cell growth normally improves resistance to lodging, pests, and disease (2).

Many other roles are attributed to K. It is required for production of high-energy phosphate (ATP) and is involved in starch as well as protein synthesis (20). It functions in N uptake and protein synthesis (3,20), lipid metabolism (22), photosynthetic processes, and carbohydrate metabolism (12).

Potassium is often described as a quality element (5,21), because fruits and vegetables grown with adequate K seem to have a longer shelf life in the grocery store.

Calcium

Calcium occurs in plants as calcium pectate, which is a component of every cell wall. It is involved in cell elongation and cell division. It influences the pH of cells and the structural stability and permeability of cell membranes. Calcium acts as a regulator ion in the translocation of carbohydrates through its effect on cells and cell walls. It plays a role in mitosis. Calcium is cited for its beneficial effect on plant vigor and stiffness of

straw and also on grain and seed formation (9). It acts as an activator for only a few enzymes. It is a part of certain structural compounds such as calcium oxalate and calcium pectate (5).

Magnesium

Magnesium is an essential part of the chlorophyll molecule. It is a cofactor for a number of enzymes including transphosphorylase, dehydrogenase, and carboxylase. Magnesium aids in the formation of sugars, oils, and fats (9). It also activates formation of polypeptide chains from amino acids (20).

Sulfur

Sulfur is a constituent of two amino acids, cysteine and methionine (some references list a third, cystine), which are essential for protein formation. It is also involved in the formation of vitamins and synthesis of some hormones. Sulfur is present in glycosides, which give the odor characteristic of onions, mustard, and garlic (20). It is a structural constituent of several coenzymes and prosthetic groups. It is also involved in oxidation-reduction reactions (14). It is a component of S-containing sulfolipids (5).

Iron

Iron is essential for the synthesis of chlorophyll. Iron is involved in N fixation, photosynthesis, and electron transfer. As an electron carrier, it is involved in oxidation-reduction reactions (9). It is also a structural component of substances involved in these reactions, such as the reduction of O_2 to H_2O during respiration (20). Iron is involved in respiratory enzyme systems as a part of cytochrome and hemoglobin and also in many other enzyme systems. It is required in protein synthesis and is a constituent of hemoprotein (9). It is also a part of iron-sulfur proteins. These act as enzymes and in electron transfer (14).

Zinc

Zinc is a metal component in a number of enzyme systems that function as part of electron transfer systems and in protein synthesis and degradation (10). Zinc is a part of auxin, one of the best-known enzymes regulating plant growth (20). Zinc reportedly can be replaced in certain enzymes by other metal elements including Co, Mn, cadmium, and lead.

Manganese

Manganese is involved in the evolution of O_2 in photosynthesis. It is a component of several enzyme systems, although in fewer than other micronutrients. It also functions in chloroplasts as part of electron transfer (oxidation-reduction) reactions and electron transport systems (9). It is a structural component of certain metalloproteins.

Copper

Copper is involved in several enzyme systems and apparently cannot be replaced by other metal ions. It is involved in cell wall formation, and like other micronutrients, in electron transport and oxidation reactions (20). Copper affects the formation and chemical composition of cell walls, which in turn affect lignification (14).

Boron

Boron is involved in the transport of sugars across cell membranes and in the synthesis of cell wall material. It influences transpiration through the control of sugar and starch formation. It also influences cell development and elongation. Boron affects carbohydrate metabolism and plays a role in amino acid formation and synthesis of proteins (11,20). It reportedly interacts with auxin (14).

Because of its impact on cell development and on sugar and starch formation and translocation, a deficiency of B retards new growth and development; hence, its deficiency symptoms will appear first as a lack of new growth (20). In soil reactions, B is similar to P in many ways. Boron in the plant does not undergo valency changes.

Dicotyledons generally require three to four times more B than do monocotyledons. Boron is absorbed as $B(OH)_3^0$ and is one of two essential elements that exist in a soil solution as an undissociated molecule.

Molybdenum

Molybdenum serves as a metal component of two enzyme systems. It is a part of nitrate reductase, which is involved in the reduction of NO_3^- to NH_4^+ after it is absorbed by the plant. It is also a structural component of nitrogenase, which is involved in the fixation of N_2 into the ammonium form in a symbiotic relationship with legumes (20).

Chlorine

Chlorine takes part in the capture and storage of light energy through its involvement in photophosphorylation reactions in photosynthesis (9). It is not present in the plant as a true metabolite but as a mobile anion (20). It is involved with K in the regulation of osmotic pressure, acting as an anion in counterbalance to cations.

Silicon

Silicon forms Si-enzyme complexes that act as protectors and regulators of photosynthesis and other enzyme activity. It plays a role in the structural rigidity of cell walls. It is absorbed as $Si(OH)_4^0$ and, like B, is taken up as an undissociated molecule. In certain plants, it is often present in percentages greater than those of N or K (20).

Sodium

Sodium is involved in osmotic regulation and will, in some cases, fulfill the function of K (20).

Vanadium

Although not considered an essential element, V is involved in plant processes. It functions in oxidation-reduction reactions, promotes chlorophyll synthesis, and has been shown to substitute for Mo (20).

Cobalt

Cobalt is involved in the growth of certain lower plant organisms. It apparently is involved in the growth of organisms involved in symbiotic N fixation (20).

Nonessential Elements

Certain nonessential elements can replace and perform the role of an essential element. Strontium reportedly replaces Ca in some fungi and algae. Tungsten has been reported to compete with Mo. Nickel is a component of the urease enzyme in certain species. Rubidium replaces K in one enzyme system (5).

The halogens—Br, I, and F—may often substitute for Cl. They are frequently found to be part of certain compounds in the plant, whereas Cl is not a part of any compound. Their functions are reported to be physicochemical, such as in the photosynthetic process; protective, as an antimetabolite; and regulatory in certain hormonal processes (11).

Diagnosing Plant Symptoms

Deficiency Symptoms

A deficiency of an element essential for plant growth will result in a decrease in the normal growth of the plant and will affect the yield of a crop. An essential element will often be sufficiently low that the deficiency will be manifested in some manner by the plant. Symptoms of nutrient deficiencies have been used for many years to diagnose growth problems.

Both essential and nonessential elements may be sufficiently in excess to be toxic to the plant and detrimentally affect growth and yield of a crop. When this occurs, the plant will usually express definitive toxicity symptoms. Such symptoms can be used to determine the source of the problem, and corrective action can be taken if it is not too late.

These symptoms can be categorized into five types: 1) chlorosis, which is a yellowing, either uniform or interveinal, of plant tissue due to a reduction in the chlorophyll formation processes; 2) necrosis, or death of plant tissue; 3) lack of new growth or terminal growth resulting in rosetting; 4) an accumulation of anthocyanin and an appearance of a reddish color; and 5) stunting or reduced growth with either normal or dark green color or yellowing.

Whether due to a deficiency or an excess, plant symptoms are often quite specific. A specific symptom, for instance, would be one due to N deficiency, which is first noted as a yellowing of the midrib of older leaves of grasses. Other less specific symptoms might indicate a deficiency of several elements. For example, a deficiency of any one of the micronutrients Fe, Zn, Mn, or Cu may cause chlorosis of younger tissue.

An example of a symptom that is often misleading is the reddish color developed in the leaves of plants. The reddish color is due to an above-normal level of anythocyanin, a red plant pigment, which accumulates when metabolic processes are disrupted. This could be caused by a deficiency of P, cool temperatures, or even maturation of the plant.

Deficiency symptoms are only one of several diagnostic tools that can be used to determine the nutrient status and nutrient needs of plants. Three other techniques commonly used include chemical soil tests, chemical plant tissue tests, and biological tests. The best method of determining nutrient need is to use two or more of the techniques. Deficiency symptoms can, at times, be quite specific in pinpointing nutrient deficiencies; however, such symptoms are often misleading because they can be caused by conditions other than nutrient deficiencies. Examples of such conditions are numerous, including herbicide damage, insect or disease infestations, root pruning, cold soil, soil that is too dry or too wet, and saline or sodic problems. A symptom typical of nutrient deficiency might even result from a hereditary condition.

Toxicity Symptoms

Certain essential plant nutrients as well as nonessential elements can be absorbed by plants in sufficient quantity to be toxic to plant growth. Essential plant nutrients that are seldom, if ever, toxic include N, P, K, Ca, Mg, S, Fe, Zn, Na, and Si. These plant nutrients, if taken up in excess, will often cause imbalances with other nutrients and will result in poor plant growth, i.e., lack of general growth, delayed maturity, and plants that are stunted or spindly.

Essential nutrients that are known to occasionally produce toxicity are Mn, Cu, B, Mo, and Cl. Toxicity symptoms are known for some of these. Manganese causes a necrosis in and a crinkling of upper leaves called crinkle-leaf in cotton (10). Excess Cl causes necrosis of the leaf tip and often a progression of necrosis down the outer edge of older leaves, a symptom quite similar to K deficiency.

Nonessential elements are often absorbed by plants. The list that can cause toxicity can be quite long. Several heavy metals are in the group and include lead (Pb), cadmium (Cd), nickel (Ni), aluminum (Al), strontium (Sr), rubidium (Rb), and tungsten (W). Symptoms due to a toxicity of these elements are seldom definitive or specific.

Marshner (14) provides an excellent review on the toxicity of nutrients and points out reasons for the occurrence of toxicities. Manganese inhibits the uptake of K and competes with the uptake of Fe, Ca, and Mg. Excess Cu competes with Fe and also inhibits root elongation and damages root cell membranes. An excess of Mo will result in the formation of molybdocatechol complexes in the vacuole, and these apparently inhibit growth processes. These complexes compete with the essential elements that are similar in valency and reactions, thereby disrupting essential metabolic processes.

Use of Deficiency Symptoms to Determine Need

Nutrient deficiency symptoms can be used to determine the nutrient needs of crops, especially when they are specific for a particular nutrient. Yellowing down the midrib and eventual dying of lower leaves is a specific indication of N deficiency, whereas chlorosis and necrosis on the outer edge of lower leaves specificly indicate K deficiency.

As discussed earlier, however, many symptoms can be caused by any one of several nutrients or by other conditions. Yellowing and intervienal chlorosis of young, newer leaves could be due to S deficiency or it could indicate a deficiency of one of the metal micronutrients, Fe, Zn, Mn, or Cu. A knowledge of soil pH and general soil conditions may be needed to determine whether the yellowing results from lack of S or one of the micronutrients. Sulfur deficiency symptoms seldom appear on high-pH soils, whereas deficiencies of Fe, Zn, Mn, or Cu seldom occur on low-pH soils.

Typical deficiency symptoms can also be caused by many other conditions. Certain herbicides, diseases, and insects can cause chlorosis in plants. Water-logged or droughty soils and mechanical or wind damage can often create problems that mimic deficiencies.

Deficiency symptoms should be used only as a guideline for determining need. Symptoms plus plant and soil analyses together with a general knowledge of crop needs and the chemistry of the soil should all be used in determining crop nutrient needs.

Determining Nutrient Needs

In a crop production program, the most difficult requirement for the grower is to estimate accurately the fertilizer needs of the crop. An effective method for estimation is outlined in a five-step program by Ulrich and his co-workers (Chapter 7, Sugar Beet) that can be used to determine nutrient need.

Start with a *visual diagnosis*. Compare unusual leaf symptoms to those in this book. Symptoms should be noted as early as they appear in the field; symptoms may be less easily classified later on because they have been modified by other factors.

Second, *verify the visual diagnosis* by comparing analytical results of leaves with and without deficiency symptoms with critical values reported in plant analysis tables for crops in this publication and in other references. Table 1.2 in this chapter also gives plant analysis guidelines. Comparable leaf samples, with and without symptoms, should be collected for chemical analysis at the same time, since the plants may outgrow the symptoms. For example, in most crops a P deficiency induced by low soil temperature may be overcome by a higher soil temperature. Also, when look-alike symptoms appear, chemical analysis can distinguish leaf scorch caused by drought or windburn from that caused by K and Mg deficiency, or even a K-deficient scorch from an Mg-deficient one. When the symptoms have been identified correctly, the analyses will be in the range that has been reported critical for most crops.

Third, *fertilize* with the required nutrients, either on a trial basis or over the entire field, leaving an unfertilized area for comparison.

Fourth, *confirm* by taking leaf samples from the field plots after rainfall or irrigation has been sufficient to ensure that the fertilizer added was actually absorbed by the plants and that the deficiency has been corrected.

Fifth, *prevent* nutrient deficiencies and crop losses in the current and succeeding crops by following a planned plant analysis program. A systematic plant analysis program can be used not only to prevent nutrient deficiencies

but also to avoid overfertilization. Adding fertilizer as insurance when the soil nutrient supply is already adequate is not only uneconomical but is often politically unacceptable when it is perceived as a possible pollution source.

Success in the five-step program will be enhanced when the physical and chemical properties of the soil are fully characterized as well as those of the fertilizer materials required to correct any deficiency from planting to harvest.

Selected References

1. Adams, F. 1980. Interactions of phosphorus with other elements in soils and plants. Pages 655-647 in: Phosphorus in Agriculture. F. E. Kwasawneh, E. C. Sample, and E. J. Kamprath, eds. American Society of Agronomy, Crop Science Society of America, and Soil Science Society of America, Madison, WI.
2. Beringer, H., and Northdurft, F. 1985. Effects of potassium on plant and cellular structures. Pages 351-364 in: Potassium in Agriculture. R. D. Munson, ed. American Society of Agronomy, Crop Science Society of America, and Soil Science Society of America, Madison, WI.
3. Blevins, D. G. 1985. Role of potassium in protein metabolism of plants. Pages 413-422 in: Potassium in Agriculture. R. D. Munson, ed. American Society of Agronomy, Crop Science Society of America, and Soil Science Society of America, Madison, WI.
4. Box, J. 1972. Potassium and crop production. Tex. Agric. Ext. Serv. Bull. MP-1007.
5. Bould, C., Hewitt, E. J., and Needham, P. 1984. Diagnosis of Mineral Disorders in Plants. Vol. 1, Principles. Chemical Publishing, New York.
6. Broyer, T. C., Carlton, A. B., Johnson, C. M., and Stout, P. R. 1954. Chlorine—A micronutrient element for higher plants. Plant Physiol. 29:526-532.
7. Buckman, H. O., and Brady, N. C. 1960. The Nature and Property of Soils. 6th ed. McMillian, New York.
8. Buckman, H. O., and Brady, N. C. 1969. The Nature and Properties of Soils. 7th ed. MacMillan, New York.
9. Follett, R. H., Murphy, L. S., and Donahue, R. L. 1981. Fertilizers and Soil Amendments. Prentice-Hall, Englewood Cliffs, NJ.
10. Foy, C. D., Webb, H. W., and Jones, J. E. 1981. Adaptation of cotton genotypes to an acid, manganese toxic soil. Agron. J. 73:107-111.
11. Gupta, U. C. 1979. Boron nutrition of crops. Adv. Agron. 31:273-303.
12. Huber, S. C. 1985. Role of potassium in photosynthesis and respiration. Pages 369-391 in: Potassium in Agriculture. R. D. Munson, ed. American Society of Agronomy, Crop Science Society of America, and Soil Science Society of America, Madison, WI.
13. Lyon, T. L., and Buckman, H. O. 1937. The Nature and Properties of Soils. 3rd ed. MacMillan, New York.
14. Marshner, H. 1986. Mineral Nutrition of Higher Plants. Academic Press, Orlando, FL.
15. Mengel, K. 1985. Potassium movement within plants and its importance in assimilate transport. Pages 397-409 in:

Table 1.2. General guidelines for critical, sufficient, and toxic levels of plant nutrients[a]

Elements	Critical level	Sufficient range	Toxicity level[b]
N, %	<2.0	2.0-5.0	Nontoxic
P, %	<0.2	0.2-0.5	Nontoxic
K, %	<1.0	1.0-5.0	Nontoxic
Ca, %	<0.1	0.1-1.0	Nontoxic
Mg, %	<0.1	0.1-0.4	Nontoxic
S, %	<0.1	0.1-0.3	Nontoxic
Fe, ppm	<50	50-250	Nontoxic
Zn, ppm	15-20	20-100	>400
Mn, ppm	10-20	20-300	>300
Cu, ppm	3-5	5-20	>20
B, ppm	<10	10-100	>100
Mo, ppm	<0.1	0.1-0.5	>0.5
Cl, %	<0.2	0.2-2.0	>2.0
Si, %	<0.2	0.2-2.0	Nontoxic
Na, %	<1.0	1.0-10	Nontoxic
Co, ppm	<0.2	0.2-0.5	>0.5
V, ppm	<0.2	0.2-0.5	>1

[a] Data are from numerous references and numerous analyses based on the author's professional experiences. Levels of nutrients in certain crops can range to higher levels without toxicities. For example, the sufficient range for S is for grains and legumes, whereas the values for crucifers are generally three to five times greater.

[b] Nutrients listed as nontoxic, when in excess, may cause imbalances and detrimentally affect growth, but they seldom are toxic.

Potassium in Agriculture. R. D. Munson, ed. American Society of Agronomy, Crop Science Society of America, and Soil Science Society of America, Madison, WI.

16. Meyer, B. S., and Anderson, D. B. 1939. Plant Physiology. D. Van Nostrand, New York.

17. Ozanne, P. G. 1980. Phosphate nutrition of plants—A general treatise. Pages 559-585 in: Role of Phosphorus in Agriculture. F. E. Kwasawneh, E. C. Sample, and E. J. Kamprath, eds. American Society of Agronomy, Crop Science Society of America, and Soil Science Society of America, Madison, WI.

18. Schrader, L. E. 1984. Functions and transformations of nitrogen in higher plants. Pages 55-66 in: Nitrogen in Crop Production. R. D. Hauck, ed. American Society of Agronomy, Crop Science Society of America, and Soil Science Society of America, Madison, WI.

19. Suelter, C. H. 1985. Role of potassium in enzyme catalysts. Pages 337-349 in: Potassium in Agriculture. R. D. Munson, ed. American Society of Agronomy, Crop Science Society of America, and Soil Science Society of America, Madison, WI.

20. Tisdale, S. L., Nelson, W. L., and Beaton, J. D. 1985. Soil: Fertility and Fertilizers. 4th ed. MacMillan, New York.

21. Usherwood, N. R. 1985. The role of potassium in crop quality. Pages 489-509 in: Potassium in Agriculture. R. D. Munson, ed. American Society of Agronomy, Crop Science Society of America, and Soil Science Society of America, Madison, WI.

22. Weber, E. J. 1985. Role of potassium in oil metabolism. Pages 425-439 in: Potassium in Agriculture. R. D. Munson, ed. American Society of Agronomy, Crop Science Society of America, and Soil Science Society of America, Madison, WI.

Part I

Grain Crops

Chapter 2

Regis D. Voss
Agronomy Department
Iowa State University, Ames

Corn

Profitable corn production depends on an adequate, but not excessive, supply of essential nutrients. The soil environment, soil pH, and soil chemical reactions determine, for the most part, the availability of nutrients to the crop. The amount of a nutrient in the harvested portion of a corn crop does not indicate its essentiality, but it does provide general information on the amount of nutrient the soil must provide or the amount to be applied over time if the soil is deficient in the nutrient. The average nutrient content in the aboveground plant material for a 160 bu/acre (10,033 kg/ha) corn crop grown in the midwestern United States is given in Table 2.1. The nutrient content of a crop depends on nutrient availability in the soil, the physical characteristics of the soil on which the crop is grown, the corn hybrid, and the growing environment for that growing season.

Nutrient deficiency symptoms are an indication that the plant cannot take up sufficient nutrients or cannot metabolize sufficient amounts of a nutrient. The reason(s) why a nutrient deficiency occurs may be complex. There may be insufficient amounts of available nutrients in the soil; applied nutrients positionally unavailable to plant roots due to patterns of root growth and soil moisture; root pruning due to machinery operations, insects, or disease; compacted soils; dry soils; or cool, wet soils. Aeration, moisture, temperature, microbial activity, and soil pH affect nutrient availability and, in some cases, the plants' ability to take up nutrients. Changing soil pH to a favorable range, which will vary for different soils but is usually from 5.5 to slightly above 7.0, will affect the availability of several nutrients. Differences in hybrids may cause nutrient deficiencies to manifest themselves in one hybrid but not in another in similar environments. Nutrient deficiency symptoms are most reliable when all other factors favor rapid growth and generally occur on several plants in a broad area following a specific soil or management pattern.

Knowing the previous crop history, native chemical and physical characteristics of the soil, current crop management practices including fertility treatments, and the weather patterns of the current year will help in arriving at a tentative diagnosis. Confirmation by soil tests and/or plant analysis (chemical tests) is needed for positive diagnosis. The appropriate plant part must be taken for plant analysis. Suggestions for plant sampling follow: 1) plants in the seedling stage to 12-in. (30-cm) tall, 15 to 30 whole plants cut off ½ in. (12 mm) aboveground; 2) plants 12-in. (30-cm) tall to tasseling, uppermost leaf from 15 to 25 plants; and 3) at early silking, ear leaf or leaf opposite and below ear from 15 to 25 plants. Sufficiency ranges of nutrients in selected parts of the corn plant are shown in Table 2.2. Nutrient deficiency symptoms are most difficult to interpret in cold, wet, or dry weather, and more than one deficiency symptom may occur under these conditions.

Deficiency Symptoms

A seed has adequate supplies of essential nutrients to support germination and seedling emergence in the absence of supplemental nutrition. The rate of germination or the rate of seedling emergence is not affected by nutrient availability. Abnormal plant growth prior to emergence is not a nutritional problem.

Table 2.1. Average amounts of nutrients in aboveground plant material in an acre of 160 bu/acre (10,033 kg/ha) corn grown in the midwestern United States

Element	Pounds (kg) of element					
	Grain		Stover		Total	
N	109	(49.5)	62.0	(28.1)	171	(77.6)
P	22.7	(10.3)	9.5	(4.3)	32.2	(14.6)
K	37.0	(16.8)	93.0	(42.2)	130	(59.0)
Ca	0.7	(0.3)	34.3	(15.6)	35.0	(15.9)
Mg	7.6	(3.5)	27.0	(12.3)	34.6	(15.7)
S	6.8	(3.1)	7.5	(3.4)	14.3	(6.5)
Fe	0.11	(0.05)	1.9	(0.9)	2.01	(0.91)
Zn	0.18	(0.08)	0.18	(0.08)	0.36	(0.16)
Mn	0.05	(0.02)	0.27	(0.12)	0.32	(0.145)
Cu	0.02	(0.01)	0.09	(0.04)	0.11	(0.05)
B	0.04	(0.02)	0.13	(0.06)	0.17	(0.08)
Mo	0.005	(0.002)	0.003	(0.001)	0.008	(0.004)
Cl	4.0	(1.8)	72.0	(32.7)	76.0	(34.5)

Plant nutritional problems rarely occur for several days after plant emergence. Normal plant growth does not occur if nutrient availability is limiting. The type of nutrient deficiency symptom that develops depends on the function of the nutrient in the plant, mobility of the nutrient within the plant, and the growth stage at which the deficiency occurs. Common deficiency symptoms are listed below:

Chlorosis—Yellowing, a common deficiency symptom that also may be caused by other environmental stresses.

Interveinal chlorosis—Commonly termed *striping*, when leaf tissue between veins turns yellow while veins remain green.

Necrosis—Commonly termed *firing*, which is complete drying and death of plant tissue. It usually begins on tips and edges of older leaves and also may be caused by drought, herbicides, disease, and foliar application of fertilizer.

Stunting—Reduced growth rate and/or shortened internodes, which may give a plant a stocky or weak, spindly appearance.

Abnormal coloration—Red, purple, brown, or abnormally dark green–reddish purple coloration caused by the pigment anthocyanin, which forms due to sugar accumulation. Anthocyanin also may form after root injury, or cool nights and warm days, or be symptomatic of the hybrid.

Location of the deficiency symptom depends primarily on the mobility of the nutrient within the plant. Where and how a nutrient deficiency is expressed depends to some extent on how easily the nutrient is translocated in the plant. Nutrients can be classified as mobile or immobile. Mobile nutrients are moved readily by the plant to young, developing tissue; and under nutrient deficiency conditions, nutrients are moved by the plant from older, developed plant parts into young tissue. Deficiency symptoms occur first and are most severe in the older leaves (base of plant), but the entire plant may develop symptoms if the deficiency is severe. Mobile nutrients are nitrogen, phosphorus, potassium, and magnesium. Immobile nutrients, under deficiency conditions, are not easily moved by the plant from older, developed plant parts into young tissue. Deficiency symptoms occur on the youngest, developing leaves, and the older leaves appear normal. Immobile nutrients are boron, calcium, copper, iron, manganese, molybdenum, sulfur, and zinc.

Identifying the type of deficiency symptom and its location on the plant and knowing some of the properties of the soil in which the plant is growing will aid in correctly diagnosing nutrient deficiencies.

Nitrogen (N)

In young corn plants, N deficiency causes the whole plant to be pale, yellowish green and have spindly stalks. V-shaped yellowing on the tips of the leaves appears later (Plate 1). Because N is a mobile nutrient in the plant, yellowing begins on the older, lower leaves and progresses up the plant if the deficiency persists (Plate 2).

Nitrogen deficiency is favored by cold or wet soil, dry soil (particularly after midseason), sandy soil, large amounts of low-N residues, large amounts of leaching rains, flooded soils, and ponded areas when the temperature is warm. Functions of N in plants include chlorophyll formation, protein formation, and carbohydrate utilization.

Phosphorus (P)

Phosphorus deficiency is usually identified on young plants. Phosphorus is readily mobilized and translocated in the plant. Plants are dark green with reddish purple tips and leaf margins (Plate 3). Phosphorus-deficient plants are smaller and grow more slowly than plants with adequate P. Some corn hybrids at early stages of growth tend to show purple colors similar to P deficiency symptoms even though P nutrition is adequate, and some corn hybrids do not show the color deficiency symptoms even though inadequate P severely limits yields. The purplish color nearly always disappears when plants grow to 3 ft or taller, but P-deficient plants will remain shorter throughout the growing season than plants with adequate P.

Phosphorus deficiency symptoms are favored by cold, too wet or too dry soils; insufficient amounts of available P in soil; positional unavailability of applied P; restricted root growth in compacted soils; and injury to roots by insects, herbicides, a cultivator, or a fertilizer side-dressing knife. Functions of P in plants include energy storage and transfer, and P is a structural component of a wide variety of biochemicals.

Potassium (K)

Potassium deficiency symptoms are first seen as a yellowing and necrosis of the leaf margins beginning on the lower leaves (Plate 4). If the deficiency persists, the leaf deficiency symptoms will progress up the plant, because K is mobile in the plant and is translocated

Table 2.2. Sufficiency ranges of nutrients in selected parts of corn plant

Element	Sufficiency ranges in plant	
	Ear leaf at silk	Whole plant, 3- to 4-leaf stage
N, %	2.7–3.5	3.5–5.0
P, %	0.2–0.4	0.4–0.8
K, %	1.7–2.5	3.5–5.0
Ca, %	0.2–1.0	0.9–1.6
Mg, %	0.2–0.6	0.3–0.8
S, %	0.1–0.3	0.2–0.3
Fe, ppm	21–250	50–300
Zn, ppm	20–70	20–50
Mn, ppm	20–150	50–160
Cu, ppm	6–20	7–20
B, ppm	4–25	7–25
Mo, ppm	0.6–1.0	...

from old to young leaves (Plate 5). Under severe K deficiency, lower corn leaves will turn yellow, but the upper leaves may remain green (Plate 6). Potassium-deficient plants tend to lodge late in the season, because these plants have lower stalk strength and are more susceptible to stalk diseases than are plants with adequate K. High N applications on a K-deficient soil can accentuate the stalk-strength problem.

These deficiency symptoms are favored by wet or compacted soil, too dry soil, sandy soil, organic soil, strongly geologically weathered soils, depletion of large amounts of K by the preceding crop, and certain tillage systems, such as ridge tillage on soils with small amounts of available K below the tillage zone. Potassium deficiency is accentuated by dry weather. Potassium in the plant is involved in enzyme activation, photosynthesis, sugar transport, protein synthesis, starch formation, and crop quality.

Calcium (Ca)

Calcium is rarely deficient in corn. Deficiency is shown when leaf tips stick to the next lower leaf, giving a ladderlike appearance (Plate 7). Plants may be severely stunted, because Ca is immobile in the plant and is not translocated from old to young plant tissue needing Ca. Low soil pH and acid soil problems are most likely to appear before Ca deficiency symptoms.

Calcium deficiencies are favored by very low soil pH (below 5.5 on mineral soils and 4.8 on organic soils) and on soils high in Mg and K. Calcium functions in the plant in cell elongation and division, structure and permeability of cell membranes, N metabolism, and carbohydrate translocation.

Magnesium (Mg)

The first sign of Mg deficiency is yellow to white interveinal striping of the leaves (Plate 8). Stripes are sometimes followed by dead round spots that give the impression of beaded streaking. The older leaves become reddish purple, and if the deficiency is severe the tips and edges may die as Mg is translocated from old to new plant tissue.

Magnesium deficiency is favored by very acid, sandy soils in regions of moderate to high rainfall where Mg has been extensively leached from the soil profile. High soil K levels or high rates of applied K in soils with low Mg levels can induce Mg deficiency. Magnesium is a constituent of chlorophyll and is involved in protein synthesis, enzyme systems, and oil synthesis.

Sulfur (S)

Symptoms in small plants are a general yellowing of the foliage as in N deficiency (Plate 9). Yellowing of the younger leaves is more pronounced with S deficiency than with N deficiency because S is not easily translocated in the plant. Stunting of plants and delayed maturity also are symptoms. Interveinal chlorosis may occur.

Sulfur deficiency is favored by acid, sandy soils; soils low in organic matter; and cold, wet soils that delay release of S from organic matter. Sulfur is involved in protein synthesis, chlorophyll synthesis, photosynthetic processes, enzyme systems, and oil formation.

Iron (Fe)

When iron is deficient, interveinal areas of upper leaves become pale green to nearly white (Plate 10). Iron is immobile and is not translocated from old to young plant tissue. Corn has a low Fe requirement, so this deficiency is rare and only occurs on alkaline soils.

Iron deficiency is favored by high-pH soils with free calcium carbonate in the surface soil and in cold, wet, poorly aerated soils. Iron is involved in chlorophyll formation, protein formation, enzyme systems, plant respiration, photosynthesis, and energy transfer.

Zinc (Zn)

Zinc deficiency symptoms vary, but they begin from the base of the leaf, extending towards the tip as interveinal light striping or a whitish band (Plate 11). The margins of the leaf, the midrib area, and the leaf tip will remain green. Plants are stunted because internodes are short. Zinc is relatively immobile in the plant, and severe Zn deficiency may result in new leaves being nearly white; this effect is termed "white bud." The nodes of Zn-deficient plants frequently show a purple to brown color when split open. Plants generally outgrow Zn deficiency unless it is severe.

Zinc deficiencies are favored by high soil pH; soil low in organic matter with high pH; and cool, wet soil. High P levels in soils that are borderline in Zn availability will often show Zn deficiencies, but high soil P levels alone do not create Zn deficiencies. Zinc is involved in protein synthesis, starch formation, enzyme systems, and growth hormones.

Manganese (Mn)

Manganese deficiency of corn is rare because corn's requirement for Mn is low. Deficiency symptoms are not clear-cut. Leaves turn olive green and may become slightly streaked (Plate 12). If deficiency is severe, leaves have elongated white streaks that turn brown in the center, deteriorate, and fall out. Manganese is relatively immobile in the plant.

High soil pH, sandy soils high in organic matter, and peat or muck soils favor Mn deficiencies. Manganese is involved in photosynthesis, enzyme systems, nitrate assimilation, iron metabolism, and chlorophyll formation.

Copper (Cu)

Copper deficiency is rare in corn. When it occurs, the youngest leaves are yellow as they come out of the whorl, and the tips may die. Copper is relatively immobile in the plant. The streaked leaves are similar to Fe-deficient leaves. The stalk is soft and limp. Some necrosis occurs on the edges of older leaves as in K deficiency.

Copper is often deficient in soils with very high organic matter and high pH. It functions in the plant in enzyme systems, protein synthesis, chlorophyll formation, and N metabolism.

Boron (B)

Boron deficiency in corn is rare. Leaves have small dead spots and are brittle (Plate 13). Boron is not readily translocated in the plant, and as a result deficient plants have a bushy appearance because upper internodes do not elongate. Tassels and ear shoots are reduced and may not emerge. Corn is very sensitive to B, and B toxicity can result if it is not applied according to recommendations (Plate 14).

Boron deficiencies are encouraged by drought, and B is usually deficient in high-pH soil and in sandy soils low in organic matter. Drought will reduce release of B from organic matter and will also delay ear shoot emergence and possibly delay pollination. Symptoms of drought and B deficiency may occur simultaneously and be confused with each other. Boron plays a role in the plant's sugar-starch balance, sugar and starch translocation, cell division, N and P metabolism, and protein formation.

Chlorine (Cl)

Symptoms of Cl deficiency have not been identified for corn. General plant symptoms are chlorosis in younger leaves and overall wilting. Factors favoring Cl deficiency have not been identified. Chlorine is ubiquitous in the environment and appears to be involved in several plant processes, including photosynthesis, sugar translocation, and maintaining or increasing leaf water potential.

Molybdenum (Mo)

Molybdenum is rarely, if ever, deficient in corn. However, if deficiency occurs, older leaves die at the tip, along the margins, and then between the veins. Molybdenum deficiency is found in soils with very low pH and in strongly geologically weathered soils. Its functions involve protein synthesis, legume N fixation, enzyme systems, and N metabolism.

Symptoms of Root Injury

Long and short strips in rows where corn either has not emerged or corn is slow growing or wilting are indicative of ammonia injury to the roots (Plate 15). The roots may be brown or turn black back to the seed (Plate 16). Corn that is slow to emerge, slow growing, or wilting as seedlings may be injured by salt from fertilizer placed too close to the seed. Roots are brown and short. Seed may be salt encrusted. Plants showing reduced growth and/or nutrient deficiencies may have root growth restricted due to soil compaction (Plate 17). Plants may also exhibit wilting, nutrient deficiencies, or growth problems from herbicide injury to the roots (Plate 18) and physiological processes of the plant.

Selected References

1. Aldrich, S. R., Scott, W. O., and Hoeft, R. G. 1986. Modern Corn Production. 3rd ed. A & L Publications, Champaign, IL.
2. Jones, J. B., Jr., Eck, H. V., and Voss, R. 1990. Plant analysis as an aid in fertilizing corn and grain sorghum. In: Soil Testing and Plant Analysis. R. L. Westerman, ed. American Society of Agronomy, Crop Science Society of America, and Soil Science Society of America, Madison, WI.
3. Tisdale, S. L., Nelson, W. L., and Beaton, J. D. 1985. Soil Fertility and Fertilizer. 4th ed. MacMillan, New York.
4. California Fertilizer Association. 1985. Western Fertilizer Handbook. 6th ed. Interstate, Danville, IL.

Chapter 3

B. R. Wells
Department of Agronomy
University of Arkansas, Fayetteville

B. A. Huey
Cooperative Extension Service
Stuttgart, Arkansas

R. J. Norman
University of Arkansas
Rice Research and Extension Center
Stuttgart, Arkansas

R. S. Helms
Cooperative Extension Service
Stuttgart, Arkansas

Rice

Paddy rice (*Oryza sativa* L.) is grown in an environment that is unique compared with growing conditions for typical upland crops. The presence of floodwater requires the rice root system to adapt to a largely anaerobic soil system. The rice plant adapts to this environment by transporting oxygen from the aerial portions of the plant to the root system (13). A secondary adaptive mechanism is an extensive system of lateral fibrous roots in the shallow zone of oxidized soil, 0.04- to 0.08-in. (1- to 2-mm) thick, located at the soil-water interface. Oxygen diffusing through the water layer allows this zone of soil to remain oxidized. For these reasons, paddy rice normally has a shallow, fibrous root system. Paddy conditions eliminate water stress as a growth factor, so the absence of a deep root system normally does not have an adverse effect on crop growth.

The aquatic environment not only influences the root system but also alters the availability of several essential elements, thus affecting nutrient uptake and utilization by the rice crop. The essential elements most affected by flooding include nitrogen, phosphorus, iron, and manganese. The availability of other elements such as potassium, zinc, calcium, and magnesium may be increased or decreased depending on soil conditions before and after flooding.

Nitrogen, phosphorus, and zinc are the three essential elements that are most often found to be limiting in rice production on a global scale; however, deficiencies or toxicities for most of the essential elements have been noted under certain conditions. Rice is also unique because it has a relatively high requirement for silicon. In leaf tissues of rice plants, silicon levels often reach 10–12% by weight.

This chapter addresses the individual essential elements in terms of their specific value to the rice plant growing in an aquatic environment and the deficiency and/or toxicity symptoms exhibited by the plants. Suggested corrective treatments will also be included where appropriate.

Deficiency Symptoms

Nitrogen (N)

Nitrogen is the essential element most often found limiting for optimum grain production by rice. Rice has a relatively high N requirement (1). In addition, the aquatic environment of the rice crop often leads to greater losses of soil N through such mechanisms as ammonia volatilization, denitrification, and leaching, resulting in a lower N uptake efficiency compared with many upland crops (4). Rice, especially at the seedling stage, prefers N in the ammonium ion (NH_4^+) form (8). This preference may be an adaptation to the aquatic environment under which most rice is grown. Nitrate (NO_3^-) N is very unstable in an anaerobic soil, usually being denitrified to N_2 gas within 3 days under temperature regimes typical for rice production. However, this preference for NH_4^+ decreases with plant age until panicle differentiation, at which time the rice plant utilizes either NH_4^+ or NO_3^- with equal effectiveness (8). At this development stage, fertilizer N applied at rates up to 30 lb/acre (33 kg/ha) of N is absorbed by rice plants within 3 days of application (12).

Nitrogen deficiency symptoms of rice include stunting, reduced tillering, reduced panicle size, and leaf chlorosis followed by necrosis (Plate 19). The older, lower leaves are most affected, since N is mobile in the plant and is readily translocated to young developing plant parts. Nitrogen deficiency also results in a slight delay in heading (1–3 days) compared with rice receiving the proper amount of N. In cases of extreme N deficiency, individual kernel weight may also be reduced (10), although this effect is relatively minor compared with the large reduction in grain number.

Nitrogen fertilizer management of rice is influenced by such factors as the soil, cultivar, and cultural system. Deep placement of an NH_4^+-N source followed by continuous flooding has been effective in the water-seeded cultural system used in California; whereas, addition of an NH_4^+-N source in split-topdressing applications timed according to plant development has been effective in the direct-seeded, delayed-flood system practiced in most of the rice-producing areas of the southern United States. In this delayed-flood cultural system, the first N application is broadcast onto dry soil when the rice is at the 4- to 5-leaf development stage; the field is then flooded. The preflood N requires approximately 3 weeks to be fully taken up by the young rice plants. Additional fertilizer N is applied into the floodwater at panicle differentiation and 10 days later (1). Both of these N management systems, if properly conducted, result in 50–75% recovery of fertilizer N by the rice crop. Nitrogen fertilizer should not be applied into the floodwater during the seedling development stage because there will be excessive losses due to NH_3 volatilization (4).

Phosphorus (P)

Phosphorus is sometimes limiting in rice production, especially on highly weathered soils with large contents of Al and Fe oxides. The reduction process that occurs in the soil following flooding normally results in an increase in P availability; therefore, on many soils no yield response to P is observed in the rice crop even though an upland crop, such as corn or cotton, growing on this soil might respond to P fertilization. Phosphorus does not undergo reduction in flooded soil; however, the ferric phosphates are reduced to the more soluble ferrous phosphates. Thus, the availability of P to rice is usually enhanced by flooding (4).

Rice, like many other grass plants, does not readily exhibit the distinct purple (anthocyanin formation) associated with P deficiency. Phosphorus deficiency symptoms are usually noted on seedling rice as severe stunting; small, very erect and dark green leaves; small-diameter stems; lack of tillering; and delayed plant development (Plate 20). These symptoms may be followed by rapid leaf necrosis (Plate 21).

On some soils where rice exhibits P deficiency symptoms during the seedling development stage, these symptoms will disappear following flooding and reduction of the ferric phosphate. Also, P-deficient rice will respond to a topdressing of P if it is followed immediately by flooding, which is possibly because of the shallow root system of the rice plant under flooded conditions. In the southern United States, where rice is rotated with upland crops such as soybeans, wheat, cotton, or corn, P fertilizer is normally applied to the other rotation crops, since they are more likely to provide an economical yield response. In California, where rice may be grown continuously for up to 10 years before rotation to an upland crop, P fertilizer must be banded rather than broadcast for the first upland crop following rice because of the extreme unavailability of the soil P to the upland crop (4). Most soil tests for P have been developed for upland crops and are not accurate for estimating the amount of P released following flooding of the soil. A combination of past history and a soil test are usually best for predicting P fertilization needs for a specific rice field.

Potassium (K)

Potassium is not involved in the oxidation-reduction process in a flooded soil; however, under most conditions, the concentration of K^+ in soil solution increases after flooding. This is generally considered to result from the increased solubility of Mn and Fe as they are reduced, move into soil solution, compete with K for sites on the exchange complex, and displace K into the soil solution. The end result is an increase of K in soil solution, increased uptake by the rice plant, and increased probability of K leaching if the soil is permeable (4).

The level of available K in the soil can be reliably determined with a soil test that estimates exchangeable K. Response of rice to direct application of K is very infrequent; thus, visual deficiency symptoms for K are seldom noted under field conditions (2). Deficiency symptoms of K on rice include stunted plants; a slight reduction in tillering; short, droopy, and dark green upper leaves; yellowing of the interveinal areas of the lower leaves, starting from the leaf tip and eventually drying to a light brown; and development of occasional brown spots on the upper leaves (Plate 22).

Potassium deficiency is readily corrected by applications of a K-containing fertilizer either to the rice crop or to another crop growing in rotation with the rice. In the southern United States, K fertilizers are normally applied to other upland crops in the rotation, such as soybeans, and the residual K from these applications is adequate to optimize rice yields.

Sulfur (S)

Sulfur deficiency has been reported in rice from various countries such as Indonesia, Brazil, India, Bangladesh, and Thailand (2). Recently, S deficiency on rice was detected in the United States in Arkansas. Sulfur deficiency is normally associated with low soil

S content that results from extreme weathering and leaching of sulfate-S. In addition, the high-analysis fertilizers being used today may contain only minute quantities of S as an impurity, and less S is being returned to the environment as an atmospheric pollutant. In Arkansas, S deficiency is confined to coarse-textured, permeable, low-organic-matter soils and sites where the irrigation water contains almost no sulfate-S. Sulfur deficiency has also been documented in Arkansas on relatively impermeable clay soils that have been cropped continuously to rice for 3–5 years. We theorize that the continuously reduced clay soils were depleted of sulfate-S during this lengthy period when the soil remained largely anaerobic.

Sulfur deficiency symptoms on rice are very similar to N deficiency symptoms, thus the two are often confused. However, S is far less mobile in the plant than N, and this will normally separate the symptoms, especially at the early stage of symptom development. This difference manifests itself as chlorosis of the upper leaves with S deficiency and of the lower leaves with N deficiency (Plate 23). In addition, the lower leaves do not become necrotic prematurely with S deficiency as is the case with deficient N. These symptoms may be noted at all stages of growth of the rice plant.

Sulfur deficiency is relatively easy to correct in rice by the application of a fertilizer containing sulfate-S. The presence of the floodwater results in immediate dissolution of the fertilizer and rapid uptake by the plants. Visual symptoms of recovery are usually noted on the rice within 5 days of fertilizer application.

Iron (Fe)

Both Fe deficiencies and toxicities have been reported in rice (13). Iron deficiency occurs on neutral to alkaline soils and is more frequent under upland conditions than in paddy. Under paddy conditions, Fe^{3+} is reduced to Fe^{2+} in anaerobic soils, and the concentration of Fe in the soil solution increases significantly. In many plants showing Fe deficiency symptoms, analysis shows a relatively high concentration of Fe in the roots and a low concentration in the shoots, indicating a problem not only with uptake but also with translocation within the plant.

With Fe deficiency, the entire leaf becomes chlorotic and is then bleached to white. Iron is immobile in the plant, so the symptoms show first on the newly emerging tissue, especially if the Fe supply to the plant is suddenly reduced (2).

Iron deficiency is very difficult to correct. Foliar sprays may be used for a temporary correction; however, these are often cost prohibitive. Usually Fe fertilizers applied to the soil are rapidly converted to insoluble forms and have very limited effectiveness. Banding $FeSO_4$ with the seed has prevented Fe chlorosis in rice grown on organic soils in Florida (5) (Plate 24). On many calcareous soils, flooding will increase the solubility of Fe and correct

the problem.

Zinc (Zn)

Rice has been shown to be sensitive to Zn, with many reports of Zn deficiency worldwide over the past 25 years. Zinc deficiency is usually associated with the seedling development stage of growth, beginning 2–3 weeks after seedling emergence, by which time the seedling has depleted the caryopsis of its stored Zn. Normally Zn deficiency in rice occurs on soils with a nonflooded pH above 7.0. Above this soil pH, Zn solubility, and thus availability, is reduced to a level at which the limited root system of the seedling rice plant cannot obtain an adequate supply from the soil solution. These elevated soil pH levels may occur naturally, or they may be induced by use of irrigation water containing appreciable quantities of calcium bicarbonates or by excessive liming of the soil. Occasionally Zn deficiency is observed on acid soils. In these instances the soil is usually highly weathered, coarse-textured, and low in organic matter.

Zinc deficiency symptoms consist of various combinations of the following: loss of turgidity of the leaves, basal chlorosis of the leaves, delay in plant development, "bronzing" of the leaves, necrosis of the leaves, and, in some cases, death of the seedling plant (Plate 25). In direct dry-seeding cultural systems, symptoms often occur shortly after flooding and are intensified in deep water (Plate 26). Bronzing is normally observed at later growth stages than loss of turgidity and leaf chlorosis. In many instances, the initial symptoms are followed by necrosis of the entire plant, especially if the floodwater is not removed. After flood removal, the remaining rice plants slowly begin recovery and may eventually produce adequate but not normal grain yields providing environmental conditions are conducive to active tillering and later to grain development. Maturity may be delayed by 3–4 weeks.

Zinc deficiency may be prevented or minimized by lowering the soil pH through the use of an acidifying agent (elemental S is the most common) or by the addition of a Zn fertilizer. Addition of a Zn fertilizer is normally the preferred choice because the cost is much lower. The suitable source and rate of Zn fertilizer are dependent on the cultural system. For instance, with transplanted rice, use of a zinc oxide dip on the roots during transplanting has been successful (2). In the water-seeded system practiced in California, a surface broadcast application of either $ZnSO_4$ or zinc lignosulfonate has been recommended, since the rice seedlings develop roots at the soil-water interface in this system. In the direct-seeded, delayed-flood system practiced in Arkansas, preplant incorporation of $ZnSO_4$, Zn chelate, or one of the Zn complexes has been successful. Also, mixing a Zn chelate with the postemergence application of the herbicide propanil has been a useful method of application. The Zn chelates allow the use of a much lower rate of Zn; however, these should not be used in a water-seeded system, because they are mobile in the soil and will move below the developing root system of the seedling rice.

Manganese (Mn)

Neither Mn deficiency nor toxicity are common in rice under field conditions. This probably results from the relatively large supply of Mn in the soil, its ease of reduction under flooded soil conditions, and the ability of the rice plant to tolerate large quantities of Mn. Under typical paddy conditions, Mn concentration begins to increase in soil solution within 3 to 5 days after flooding and remains at elevated levels throughout the flood period of the crop. Rice can tolerate tissue levels of more than 2,500 ppm of Mn without adverse effects on either growth or grain yield (7).

Manganese deficiency causes stunting of the plant and interveinal chlorosis of the new leaves but does not have any affect on tillering (Plate 27). These symptoms are most likely to appear on rice growing on high-pH soils under nonflooded conditions or on Histosols naturally low in Mn. Addition of a flood with the resultant reduction of the soil and release of reduced Mn will normally correct the problem.

Silicon (Si)

Silicon has not definitely been shown to be an essential element for rice (13). Although the essentiality of Si has not been proven, Si application on organic soils has significantly increased Si concentration in the leaves and grain yields (6). Silicon also benefits rice in several other ways, including improved photosynthetic activity, improvement in water use efficiency, increased resistance to disease (fungal) and insect damage, increased straw strength and leaf turgor, improved P metabolism, increased grain and milling yields, and improvement of overall plant nutrition (13).

Symptoms of Si deficiency are soft, droopy leaves; increased lodging; reduction in grain yield; and increased incidence of brown spot, resulting from infestation by *Helminthosporium oryzae* Breda de Haan (3) (Plate 28). Silicon deficiencies normally occur on low-Si, organic soils or on highly weathered soils that have been depleted of Si. Application of Si-containing fertilizers such as furnace, dolomite, and calcium silicate slags will usually correct Si deficiency.

Toxicities

Elemental toxicities of rice are normally a localized problem. Iron toxicity has been reported on upland rice growing on highly weathered, extremely acid soil and on paddy rice growing on acid sulfate soils. Rice can tolerate large quantities of Mn; hence, Mn toxicity is not a problem. The most common toxicity problems for rice growing in the United States are salinity and straighthead.

Salinity

Soil salinity is a leading cause of reductions in rice yield at many locations throughout the rice-producing areas of the world. Sodium chloride is the predominant salt; however, salts of Ca or Mg may be present. The first effect on the plant is normally osmotic, but if the salt levels are sufficiently high, toxic ionic effects may also be important. Generally, problem soils of this type are either saline or sodic. The electrical conductivity of the soil solution for saline soils is greater than 4 dS/m at 75°F (25°C). The sodic soils have an exchangeable sodium percentage on the cation exchange complex of more than 15%.

Rice is most susceptible to salinity at the seedling development stage. If the seedling survives the salinity conditions, rice becomes more tolerant until anthesis, when it is again susceptible. At the seedling stage, the symptoms of excess salinity include whitish leaf tips and leaf tip dieback, stunting, reduced tillering, and death of the plant (Plate 29). Salinity damage at anthesis is manifested by an increase in floret sterility. Rice growing on sodic soil is normally characterized by poor germination and emergence, severe stunting, decreased tillering, and lack of response to any nutritional additions or other external factors.

Straighthead

Straighthead is a physiological problem noted on rice growing on certain soils in the southern United States. The rice plant has normal growth until the start of the reproductive stage, at which time either panicles are distorted or they do not form.

Straighthead symptoms consist of distorted panicles with missing or blank florets and misshapen grains termed parrot beaking (Plate 30). The rice leaves remain dark green and maturity is delayed. This problem may be induced by residues of arsenical herbicides in the soil (Plate 31); however, it also occurs on soils without a history of arsenical herbicides (9). Considerable diversity among genetic lines exists for susceptibility to this problem; therefore, selection of a cultivar with resistance to straighthead is the preferred method of control. Removal of the flood and drying (oxidizing) the soil just prior to panicle initiation will also minimize the damage from straighthead.

Critical and Sufficient Nutrient Levels for Rice

Plant and soil analyses are often required to confirm the causes of visual symptoms. Although affected by climatic and other variables, concentrations of elements in rice plant tissue can be a valuable tool in confirming both deficiencies and toxicities. Use of plant tissue concentrations in diagnosing nutritional problems requires a knowledge of the plant part to sample, the stage of plant development, and the general growing conditions at the time of sampling.

Nitrogen

For diagnostic purposes, N concentration is usually determined on the most recently matured rice leaf (Y leaf) (1), with the critical concentration being highly dependent on the stage of plant development. A critical range of 2.5–3.2% N in the Y leaf at panicle initiation has been advocated by researchers in Louisiana and California.

Phosphorus

Phosphorus concentration varies less with plant development than N does. Generally a concentration above 0.1% in the leaf blade during the tillering stage of development is considered adequate for normal growth.

Potassium

Potassium, like P, does not vary greatly with stage of plant development. A concentration of 1.0% or more in the leaf blade during the tillering stage of plant development is considered adequate.

Sulfur

The critical S concentration in rice tissue, like N, varies with the stage of plant development and portion of the plant being sampled. The S concentration should be determined on the Y leaf of the plant. The critical concentration varies from approximately 0.25% at tillering to 0.1% at heading.

Iron

Rice has a relatively wide range in plant tissue Fe concentration between deficiency and toxicity. Iron deficiency at the tillering stage occurs with Fe leaf blade tissue concentrations of less than 70 ppm; at the same stage of development, toxicity occurs at concentrations greater than 300 ppm.

Zinc

Zinc deficiency in rice is associated with the seedling stage of plant development, so most research on tissue Zn concentration has emphasized this growth stage. Zinc concentrations in the leaf blade of less than 10 ppm are a definite sign of deficiency, whereas concentrations between 10 and 15 ppm indicate possible deficiency. Care in sampling is important because careful washing of the plant sample is required to remove soil contaminants from the small plants. Rice is very tolerant of Zn, with toxicity levels being estimated at more than 1,500 ppm.

Manganese

Rice is also very tolerant to high levels of Mn, with deficiencies observed at shoot concentrations of less than 20 ppm and toxicities at more than 2,500 ppm.

Silicon

Rice responds to Si and often exhibits deficiency symptoms when the Si concentration in the leaf or stem drops below 5%.

Selected References

1. Brandon, D. M., and Wells, B. R. 1986. Improving nitrogen fertilization in mechanized rice culture. Ch. 9 in: Nitrogen Economy of Flooded Rice Soils. W. H. Patrick, Jr., and S. K. DeDatta, eds. Martinus Nijhoff, Dordrecht, The Netherlands.
2. DeDatta, S. K. 1981. Principles and Practices of Rice Production. John Wiley & Sons, New York.
3. Elwad, S. H., and Green, V. E., Jr. 1979. Silicon and the rice plant environment: A review of recent work. Riso 28:235-253.
4. Patrick, W. H., Jr., Mikkelsen, D. S., and Wells, B. R. 1986. Plant nutrient behavior in flooded soils. Pages 197-228 in: Fertilizer Technology and Use. 3rd ed. O. P. Englested, ed. Soil Science Society of America, Madison, WI.
5. Snyder, G. H., and Jones, D. B. 1988. Post emergence correction of iron related rice seedling chlorosis in the Everglades. Proc. Rice Tech. Working Group 22:51.
6. Snyder, G. H., Jones, D. B., and Gascho, G. J. 1986. Silicon fertilization of rice on Everglades Histosols. Soil Sci. Soc. Am. J. 50:1259-1263.
7. Tanaka, A., and Yoshita, S. 1970. Nutritional Disorders of the Rice Plant in Asia. Int. Rice Res. Inst. Tech. Bull. 10.
8. Wells, B. R. 1960. Nitrogen absorption by rice as determined by the use of the stable isotope ^{15}N. M.S. thesis. University of Arkansas, Fayetteville.
9. Wells, B. R., and Gilmour, J. T. 1977. Sterility in rice cultivars as influenced by MSMA rate and water management. Agron. J. 69:451-454.
10. Wells, B. R., and Turner, F. T. 1984. Nitrogen use in flooded soils. Pages 349-362 in: Nitrogen in Crop Production. R. D. Hauck, ed. American Society of Agronomy, Madison, WI.
11. Wells, B. R., Norman, R. J., Lee, F. N., and Moldenhauer, K. A. K. 1988. Integrating management practices to maximize rice yields. J. Fert. Issues 5(1):14-18.
12. Wilson, C. E., Jr., Norman, R. J., and Wells, B. R. 1989. Seasonal uptake patterns of fertilizer nitrogen applied in split applications. Soil Sci. Soc. Am. J. 53:1884-1887.
13. Yoshita, S. 1981. Fundamentals of Rice Crop Science. International Rice Research Institute, Los Baños, Philippines.

Chapter 4

R. B. Clark
USDA-ARS, Department of Agronomy
University of Nebraska, Lincoln

Sorghum

Nutrient Requirements

Nutrients required for sorghum growth and production are determined by available and residual nutrients in soils, soil chemical and physical properties, environmental conditions of soils, and desired yield. Some of these factors can be controlled by man, whereas others cannot. When nutrients are inadequate, in excess, or imbalanced, sorghum plants will normally develop deficiency or toxicity symptoms. Most nutrient deficiencies have specific and unique symptoms, whereas some nutrients or elements in excess show symptoms unique to the element in excess, and some may cause other nutrient deficiency symptoms to appear.

Element deficiency or toxicity symptoms are often associated with tissue concentrations. Optimum nutrient concentrations in sorghum plants vary with plant age, stage of development, and plant part. Young vegetative plants usually have higher concentrations of mineral nutrients than older plants, and stalks or grain have different concentrations from leaves. Even leaves at different positions on the stalk and areas of the leaf (margin, vein, tip, base, or midrib) vary in nutrient concentrations. Some nutrients are transferred or remobilized from one plant part to another as plants develop and mature while other nutrients are not. These factors should be considered when determining critical and excess concentrations in plants associated with deficiencies or toxicities. Table 4.1 shows nutrient concentrations noted in field-grown sorghum by age and plant part under relatively optimum conditions. Table 4.2 shows concentrations of mineral nutrients in sorghum considered to be deficient, marginal, or critical; high; and toxic. Concentrations listed in Table 4.2 are given as guidelines, because concentrations may vary depending on experimental and growth conditions such as site of growth (field, greenhouse, growth chamber), growth media used

(soil, sand, nutrient solutions), stage of growth, plant part, and leaf used for analysis. For discussion of various nutrient deficiency symptoms, these values may be helpful for better understanding accumulation and distribution of nutrients in sorghum by plant part and age. Additional information on nutrient concentration ranges and critical levels in sorghum is presented in the literature (4–6).

Nutrient Deficiency Symptoms

Some visual symptoms and descriptions of nutrient deficiency and toxicity symptoms in sorghum have been reported (1–4).

Nitrogen (N)

Plants deficient in N are usually stunted, spindly, and pale green to pale yellow. Because N is readily remobilized from older to younger tissues, deficiency symptoms appear first on older (lower) leaves and advance to younger (upper) leaves (Plate 32). A fairly uniform pale or deep yellow color develops near the tips and margins and progresses toward the base and midrib of the leaf. The boundary between affected and unaffected tissue is usually diffuse. Dark brown necrotic spots often develop when severe N deficiency occurs. Severely N-deficient leaves turn pale brown, die, and bend over or fall down (pendent) on the plant. Nitrogen-deficient plants grow more slowly, have delayed flowering, are smaller in size, have reduced yields, and usually have fewer seeds.

Phosphorus (P)

Phosphorus deficiency is common during cool weather in young sorghum plants. Shoot growth is generally affected more by P deficiency than roots, so shoot-root ratios normally decrease as plants become more deficient. Grain development and filling are inhibited by P deficiency, so kernels are often shriveled and of poor quality. Deficiency of P is characterized by stunted, spindly plants with low vigor and dark green leaves,

Author's present address: Appalachian Soil and Water Conservation Resource Laboratory, Agricultural Research Service, Beckley, West Virginia.

which have overtones of dark red coloration. Because P is readily remobilized, older sheaths and leaves first show red pigmentation, which progresses to younger leaves (Plate 33). Patterns and sharpness of red pigmentation may be used to distinguish P deficiency from the natural red pigmentation that often appears on leaves of many sorghum genotypes. Phosphorus-deficient leaf tips and interveinal tissue show redness that progresses toward the base, veinal tissue, and midrib; eventually, the whole leaf is covered with a uniform red color. Boundaries between affected and unaffected tissue are usually distinct. If the deficiency continues, leaves turn pale brown and die. Leaves of P-deficient young plants often appear more erect and sometimes "leathery." Roots will normally turn dark brown, purple, or black.

Potassium (K)

Potassium is readily remobilized from older to younger tissue, thus K deficiency symptoms usually appear first on older leaves and progress to younger leaves as the severity of the deficiency develops. However, some sorghum genotypes show K deficiency symptoms initially on middle or relatively young mature leaves that later spread to older and developing leaves (Plate 34). Irregular necrotic patterns intermingled with red or brown pigmentation characterize K deficiency symptoms. Streaked patterns sometimes appear on interveinal tissue, but the symptoms are fairly uniform over the leaf. Symptoms begin at leaf tips and margins and move toward the base and midrib. The necrotic interveinal lesions near the margins sometimes extend for most of the leaf length. Unaffected portions of the leaf remain green. Potassium deficiency may be difficult to distinguish from "red-speckling" caused by excess P and some Mg deficiency symptoms (Plate 35). Spindly growth is not normally observed on K-deficient compared with

N- or P-deficient plants. Shoot-root ratios of K-deficient plants remain fairly constant, and grain yields are frequently reduced.

Calcium (Ca)

Mild or temporary Ca deficiency symptoms are often noted as serrated (torn and warped) upper leaves whose margins normally show irregular bleached or pale yellow-green streaking (Plate 36). These leaves are usually short, erect, and easily torn to show a sawtoothed (serrated) appearance. Plants with severe Ca deficiency are stunted since internodes fail to elongate and the new growth grows in a rosette form. Leaf tips fail to unfold, are deformed, and form swordlike projections. These tips sometimes envelop other leaves trying to emerge from the whorl to form a ladderlike appearance with blade tips sticking together. Leaves become brittle, frequently coalesce, turn brown, and form sticky vesicles at or near the margins. Calcium deficiencies appear in newly emerging leaves, and meristems are often destroyed because Ca is not easily remobilized from lower to upper leaves. In severely Ca-deficient plants, differentiation between affected and unaffected tissue is usually distinct; adjacent unaffected tissue is normally dark green, and tissue just undergoing Ca deficiency may appear water-soaked. Enhanced tillering often occurs in plants with severe Ca deficiency because apical meristems are destroyed. Flowering and maturity are delayed by mild Ca deficiency. Shoot-root ratios usually decrease because shoots are affected more extensively than roots. Calcium deficiencies are common in greenhouse- or growth-chamber-grown sorghum plants because of reduced transpiration, restricted root growth, high light intensities and/or light from certain kinds of lamps (sodium-vapor), high temperatures, and source of N (ammonium-N and urea enhance Ca deficiency

Table 4.1. Nutrient concentrations (dry weight basis) in aboveground sorghum plant parts

| | | Stage of plant growth | | | | | | | |
| | | Bloom/early grain fill | | | | Maturity | | | |
Parameter or nutrient	Vegetative[a] (whole plant)	Leaves	Stalk	Panicle	Whole plant	Leaves	Stalk + panicle	Grain	Whole plant
Dry matter yield, g/plant	58	80	131	13	224	88	106	249	443
N, %	2.04	2.86	1.10	2.33	1.79	1.96	0.60	1.37	1.30
P, %	0.32	0.31	0.24	0.42	0.28	0.18	0.11	0.30	0.23
K, %	4.02	1.94	2.08	0.89	1.98	1.00	2.46	0.57	1.11
Ca, %	2.86	4.11	2.32	0.55	2.86	5.84	2.53	0.13	1.84
Mg, %	0.22	0.25	0.29	0.34	0.28	0.23	0.26	0.22	0.23
S, %	0.14	0.14	0.09	0.20	0.11	0.12	0.09	0.10	0.10
Si, %	2.18	2.38	1.73	0.10	1.88	4.12	2.66	0.03	1.47
Mn, ppm	41	40	40	29	40	73	25	11	26
Fe, ppm	274	196	62	41	108	393	67	33	112
Cu, ppm	7.4	8.4	7.9	10.7	8.4	7.9	7.5	4.2	5.7
Zn, ppm	21	20	27	39	26	23	21	14	18

[a]12-Leaf stage.

symptoms more than nitrate-N). Sorghum genotypes differ extensively in susceptibility to Ca deficiency and expression of symptoms.

Magnesium (Mg)

Magnesium-deficient plants lack vigor, are often stunted, and usually have a delayed reproductive stage. Deficiency symptoms appear first on older leaves and progress to younger leaves (Plate 37). Relatively large irregular necrotic spots or lesions appear uniformly on leaf tips and margins and progress toward the base and midrib. Small lesions often connect with others, causing long, wide, necrotic lesions. Plants often turn pale green, and leaves may be streaked because of greater yellowing in the interveinal tissue. Leaf color in Mg-deficient plants varies considerably with sorghum genotype and may appear dark green (except where lesions occur), pale yellow green, orange, light red, and brown. Leaves with Mg deficiency are often difficult to distinguish from leaves that are K-deficient or red-speckled from excess P (Plate 35). Severely Mg-deficient leaves may become brittle, then die and turn brown. Shoot-root ratios increase with Mg deficiency because root growth decreases more than shoot growth.

Sulfur (S)

Sulfur-deficient plants are stunted and less vigorous than healthy plants and normally have decreased growth and reduced grain yield. Symptoms appear first in upper leaves, which turn pale yellow, and are more pronounced near the whorl, with enhanced greening towards the tips (Plate 38). Although the pale yellow color is fairly diffuse, light streaking along the interveinal tissue normally appears. Emerging leaves often turn uniformly pale yellow. Except for the position of affected leaves (upper instead of lower), early S deficiency is often indistinguishable from N deficiency. Another set of symptoms may appear on older leaves under prolonged S deficiency; leaves turn pale green with a chlorosis developing first near the tip and advancing along the margin toward the base. Eventually chlorotic tissue turns brown, and leaves die and hang down the stem. Shoot-root ratios usually decrease for plants grown under S deficiency.

Manganese (Mn)

Manganese deficiency normally appears first on middle to upper leaves as distinct lesions in the interveinal tissue near the middle of the leaf and progresses toward the tip (Plate 39). Sometimes these lesions will extend almost the full length of the leaf. Long chlorotic streaks may be interspersed with smaller dark brown or red lesions that join to form a long, dark brown or red lesion. Similar symptoms may also appear on leaves emerging from the whorl, with the most severe symptoms appearing on leaf portions nearest the whorl. Under mild deficiency conditions, the vein usually remains green with distinct, long interveinal lesions.

However, the lesions may diffuse and cross the veins under severe deficiencies, causing nearly the entire leaf to be affected. Manganese deficiency has about the same coloration as K and Mg deficiency and red-speckling from excess P, but there are distinct differences in symptoms or lesions (see Plate 35). Mild Mn deficiencies seldom affect flowering and maturity, but severe deficiencies can cause plants to die. Shoot-root ratios tend to increase in Mn-deficient plants.

Iron (Fe)

Sorghum is known for its susceptibility to Fe deficiency chlorosis (often called lime-induced chlorosis) when grown on alkaline calcareous soils. Iron deficiency chlorosis is common in whole fields or large areas within fields. Symptoms appear first on young leaves, with long yellow streaks in the interveinal tissue and leaf tips and margins normally more yellow than the midrib or base (Plate 40). Leaves may become even more yellow and turn a bleached white under severe deficiency conditions, with leaf tips and margins eventually turning brown and

Table 4.2. Concentrations of nutrients in sorghum[a]

| Nutrient | Growth stage | Concentration[b] | | |
		Deficient, marginal, or critical	High	Toxic
N, %	Seedling	3.3–5.3	...	...
	Vegetative	2.5–3.2	>4.0	...
	Flowering	2.0–3.0	...	...
	Grain filling	1.6–1.8	...	...
P, %	Seedling	0.10–0.25	>0.5	...
	Vegetative	0.07–0.17	...	...
K, %	Seedling	<2.0	...	...
	Grain filling	1.0–1.7	>4.0	...
Ca, %	Seedling	0.2–0.4	>6.0	...
Mg, %	Seedling	<0.2	>0.8	...
	Grain filling	0.08–0.12	>0.5	...
S, %	Seedling	0.20–0.27	...	...
	Grain filling	<0.15	...	...
Cl, %	Seedling	...	...	>0.2
Na, ppm	Seedling	...	...	>30
Mn, ppm	Seedling	10–15	>200	>500
Fe, ppm	Seedling	50–100	>400	...
Zn, ppm	Seedling	9–15	>150	>300
Cu, ppm	Seedling	2–5	>50	...
B, ppm	Seedling	3–7	>35	...
Al, ppm	Seedling	...	>200	...

[a] Values are from various sources (4,5). Tissues analyzed were leaves or whole plants in the seedling and vegetative growth stages and leaves in the flowering or grain-filling stages.
[b] Percentages are grams per kilogram $\times$ 10; parts per million are milligrams per kilogram.

dying. The color border between the vein and interveinal tissue is distinct in mildly chlorotic leaves but becomes more diffuse in severely chlorotic leaves. Plants treated with Fe can show marked differences in color compared with adjacent untreated plants (Plate 41). Even without Fe treatments, sorghum normally regreens as the growing season progresses. Few effects on growth and yield occur on plants with a slight Fe deficiency, but severely deficient plants have reduced growth and grain yields with delayed flowering and maturity dates.

Zinc (Zn)

Zinc deficiencies appear first on young leaves but are most pronounced later after leaves unroll. Leaves near the whorl show elongated pale green to bleached white sections that lengthen and become more pronounced to cover much of the leaf when deficiencies become severe (Plate 42). Sorghum genotypes vary in broadness of chlorotic bands and may show reddish purple pigmentation, especially on the underside of affected leaves, or red-brown sections next to or over the bleached portion of the leaf. Sometimes a narrow red band appears along the margin. The boundary between affected and unaffected tissue is fairly diffuse. Zinc-deficient plants have shortened internodes (stunted), and flowering and maturity are delayed. Shoot-root ratios of Zn-deficient plants normally decrease slightly.

Copper (Cu)

Copper deficiency appears first on leaves within the whorl and on young mature leaves. Mild Cu deficiency turns leaves pale green yellow primarily in interveinal tissue. Expanding leaves emerging from the whorl may remain tightly curled. Pale brown necrosis develops near the tips of younger leaves, and leaf tips and margins wither and die to form a twisted spiral (Plate 43). The twisted brown tip usually hangs down from the green basal tissue of the leaf. Leaf tips of very severely Cu-deficient plants may resemble those affected by Ca deficiency when tissue remains stuck together as leaves emerge from the whorl. Copper-deficient plants are stunted and growth is depressed.

Boron (B)

Boron deficiency can stunt plants severely by shortening internodes (Plate 44). Leaves are normally dark green, short, erect, brittle, and easily torn or broken. White or transparent lesions, which darken with time, often develop intermittently in the interveinal tissue at the midsection of leaves emerging from the whorl. These lesions may extend laterally beyond the midsection and join other lesions when plants have severe B deficiency. The boundary between affected and unaffected tissue is diffuse. Severe B deficiency destroys apical meristems, and tillers develop. If severe B deficiency develops before the boot stage of growth, tillers may appear from nodes well above the crown.

Other Nutrients

Although molybdenum, silicon, low aluminum, sodium, chlorine, and other elements may benefit sorghum growth, none of these have been shown to be essential for this plant species. No deficiencies have been detected when plants were grown without these elements.

Toxicity Symptoms

Any nutrient or element administered in sufficiently high concentrations will be toxic to plants. Toxicity symptoms may be due to the excess nutrient and its carrier or to imbalances of other nutrients. In a greenhouse study of sorghum grown in nutrient solutions, excess amounts of the various nutrients and certain elements were administered to obtain toxicity symptoms. Concentrations of these nutrients or elements were not determined.

Nitrate-N

Excess nitrate-N administered as calcium nitrate caused leaves to be lighter in color with reddish purple lesions extending throughout the leaf and interveinal tissue turned yellowish brown (Plate 45). Severely affected leaves had necrotic spots intermingled with red and yellow-brown areas. Excess nitrate-N reduced root growth, especially secondary roots, and increased fibrous root hair growth.

Ammonium-N

Leaves of plants grown with excess ammonium-N applied as ammonium nitrate formed dark reddish brown lesions near the margins and had lighter colored leaves (Plate 46). Under severe conditions, the margins turned brown and died.

Phosphorus

Many sorghum genotypes grown with relatively low levels of P developed reddish purple speckles (red-speckling) and lesions on the lower leaves of greater intensity at the margins and near the tip than at the midrib or base (Plate 47). Leaf red-speckling intensity diminished as it progressed toward upper leaves (Plate 47). Intensity of red-speckling increased when higher levels of P were administered. Leaves turned dark reddish brown and died as red-speckling became more severe. If sufficiently high levels of P are administered, upper leaves may become Fe deficient in addition to red-speckling on lower leaves. Severe P toxicity red-speckling symptoms are difficult to differentiate from Mg and K deficiencies (see Plate 35). Sorghum genotypes differ in susceptibility to red-speckling, which can occur on sorghum leaves at P levels considered normal for optimum plant growth; calling the red-speckling P toxicity seems almost inappropriate. Nevertheless, the symptom readily develops. In fact, studies with low P levels showed red-speckling on lower leaves of sorghum that 3 days later became P deficient. Sorghum appears to be unique among the cereals in developing red-speckling from added P.

Potassium

Excess K as potassium chloride caused "firing" and severe browning of leaf tips. Symptoms progressed uniformly toward the base (Plate 48), and leaves wilted, shriveled, and died. Sharp boundaries were noted between affected and unaffected tissue. Plants given excess K as potassium sulfate showed less severe burning and some reddening compared with plants grown with potassium chloride.

Calcium

Leaves of plants grown with excess Ca as calcium chloride showed red spots and streaks, with greater intensity near the tip and a lighter than normal color over most of the leaf (Plate 49). Leaves wilted and died. Boundary differences between affected and unaffected tissue were diffuse. Symptoms were less severe from Ca added as calcium sulfate than comparable levels of calcium chloride.

Magnesium

Excess Mg given as magnesium chloride caused leaf tips to turn brown and die, and interveinal leaf tissue was lighter near the tip than the base (Plate 50).

Manganese

Manganese toxicity is a common disorder for sorghum grown on many acid soils (pH ≤ 4.5 to 5.0). Manganese toxicity symptoms on leaves normally consist of dark green leaves with large numbers of small, dark, reddish purple spots distributed somewhat uniformly over the leaf. At more severe toxicity conditions, these dots may join together, leaving dark red streaks in the interveinal tissue of the leaf (Plate 51). These symptoms appear readily on older leaves and advance to younger leaves. Another common symptom of excess Mn is interveinal yellow streaking, typical of Fe deficiency, in young leaves emerging from the whorl. The leaf base is affected more than the tip.

Iron

Iron toxicity can be a problem for sorghum grown on some acid soils. Excess Fe given to plants caused leaves to turn light with blackish, straw-colored lesions at the margin (Plate 52). The boundary between affected and unaffected tissue was relatively diffuse. Older leaves were affected first, and symptoms advanced to the younger leaves.

Zinc

Margins and tips of leaves from plants grown with excess Zn turned a lighter color with necrotic lesions in the interveinal tissue. The chlorosis spread uniformly over most of the leaf. Some water-soaking symptoms appeared at tips, and symptoms typical of Fe deficiency appeared in leaves emerging from the whorl. The boundaries between affected and unaffected tissue were diffuse.

Copper

Leaves from plants given excess Cu turned lighter in the interveinal tissue, giving streaks similar to Fe deficiency with red streaks along the margins.

Boron

Excess B caused leaf margins and tips to turn straw brown, with sharp boundaries between affected and dark green unaffected leaf tissue (Plate 53).

Molybdenum (Mo) and Selenium (Se)

Symptoms on leaves from plants given excess Mo and Se were indistinguishable from P deficiency symptoms (Plates 54 and 55); however, excess Se caused interveinal yellow streaking.

Salinity: Sodium (Na), Chlorine (Cl), and Sulfur

Saline conditions caused by addition of Na, Cl, or sulfate can give similar toxicity symptoms on leaves. Addition of the osmotic agent polyethylene glycol-6000 also caused similar symptoms. Plants were stunted, and older leaves were affected more than younger leaves, although both were affected extensively when excess Na was added. Leaves became flaccid and wilted, and margins rolled upwards towards the tip. Margins and tips were initially gray and then turned brown and died, leaving green midveins. Boundaries between affected and unaffected tissue were distinct. Excess Cl also caused red lesions on leaves.

Aluminum (Al)

Aluminum toxicity is a major problem for sorghum grown on acid soils with a pH of less than 5.0 if genotypes are not tolerant to the conditions (Plate 56). Plant growth is reduced, and plants have a light, diffuse chlorosis, which is more severe on lower leaves and advances to upper leaves. Tips and margins of leaves normally turn brown and die, with symptoms advancing from tip to base. Since Al toxicity affects roots more dramatically than shoots, uptake of nutrients and water are often restricted, thus enhancing deficiencies of many other nutrients. Genotypes vary in Al toxicity symptoms, but symptoms typical of Fe deficiency are common, and P, Ca, and Mg deficiencies may occur. Toxicity of Al reduces root elongation and causes roots, especially tips, to swell, become stubby, coralloid, brittle, and dark.

Silicon (Si)

Silicon toxicity symptoms were noted only if high levels of Si were added to sorghum plants. Silicon can alleviate Mn and Al toxicity symptoms. Nonuniform, splotchy yellow sections develop on margins of leaves and become uniformly light yellow, with some red splotching over most of the leaf as severity increases (Plate 57). If excess levels of Al or Mn are also included in solutions, the severity of Si toxicity can be reduced.

Barium (Ba) and Strontium (Sr)

Excess Ba caused dark red lesions near the midrib and lighter leaf color near leaf margins. Symptoms were more severe at the leaf base, coming from the whorl, than they were at the tip. Necrotic lesions in a splotchy pattern and a lighter color developed first at the margins and progressed toward the midrib on leaves of plants given excess Sr (Plate 58).

Cobalt (Co) and Nickel (Ni)

Yellow streaking in the interveinal tissue typical of Fe deficiency appeared at the base of leaves emerging from the whorl when excess Co (Plates 59) and Ni (Plate 60) were added. Leaf sheaths of plants given excess Co also had yellow streaking. The symptoms did not extend to leaf tips.

Chromium (Cr), Lead (Pb), and Cadmium (Cd)

Leaves of plants given excess Cr developed a light, reddish brown color from the tip towards the base and from the margin towards the midrib (Plate 61). Relatively dark reddening was noted at leaf tips and margins, and the boundary between affected and unaffected tissue was fairly diffuse. Excess Pb given to plants caused relatively bright reddish brown symptoms, which appeared first on leaf tips and margins and progressed toward the base and midrib (Plate 62). The boundary between affected and unaffected tissue was distinct. A relatively bright reddish orange color developed on margins and tips of old leaves of plants grown with excess Cd. Symptoms advanced toward the base and midrib, and brilliant red leaves of uniform color appeared (Plate 63) as Cd toxicity symptoms became severe.

Selected References

1. Clark, R. B. 1982. Plant response to mineral element toxicity and deficiency. Pages 71-142 in: Breeding Plants for Less Favorable Environments. M. N. Christiansen and C. F. Lewis, eds. John Wiley & Sons, New York.
2. Clark, R. B. 1988. Mineral nutrient requirements and deficiency/excess disorders of sorghum. Crop Res. 1:16-35.
3. Frederiksen, R. A. 1986. Compendium of Sorghum Diseases. American Phytopathological Society, St. Paul, MN.
4. Grundon, N. J., Edwards, D. G., Takkar, P. N., Asher, C. J., and Clark, R. B. 1987. Nutritional Disorders of Grain Sorghum. Australian Centre for International Agricultural Research, Canberra.
5. Reuter, D. J. 1986. Temperate and sub-tropical crops. Pages 39-99 in: Plant Analysis: An Interpretation Manual. D. J. Reuter and J. B. Robinson, eds. Inkata Press, Melbourne, Australia.
6. Westerman, R. L. 1990. Soil Testing and Plant Analysis. 3rd ed. Soil Science Society of America, Madison, WI.

Chapter 5

M. V. Wiese
Plant Pathology Division/PSES
University of Idaho, Moscow

Wheat and Other Small Grains

Wheat and small grains and all other agricultural crops require an abundant supply of nutrients for optimal growth and reproduction. The largest requirements are for carbon (C), hydrogen (H), and oxygen (O). These elements, derived mainly from air and water, form the bulk of all plant proteins, carbohydrates, and fats, and the bulk of all structural tissues. The supply of C, H, and O in water and air is abundant, relatively fixed, and unmanageable. On the other hand, 13 other nutrient elements (N, P, K, Ca, Mg, S, Fe, Zn, Mn, Cu, B, Cl, and Mo) are present in the soil, wherein they can vary in quantity and in availability. These soil nutrients are manageable; and because they are consequential when in short supply and sometimes when they are in excess, they are the subject of this discussion.

In essence, the production of wheat and other small-grain cereals is a mining operation where essential nutrient elements are removed from the soil, taken into the developing plant, and removed in part at harvest. Harvest of these soilborne nutrients, and their incomplete replacement or their loss through soil leaching and erosion lead to their eventual depletion. Thus, nutrient supplementation of soil, especially through application of chemical fertilizers and organic matter such as manure, is commonplace and usually necessary to attain economic yields of small-grain cereals.

Nutrient deficiencies and/or toxicities in wheat and other small grains are often subtle and difficult to detect. Most nutrient deficiencies inflict significant damage and loss before deficiency symptoms appear. With time, nutrient disorders will produce visible symptoms on affected plants (Plate 64). For example, yellowing of younger tissues coupled with a quick sap test for N can offer conclusive evidence of N deficiency. In other instances, discoloration of leaves and other plant parts in shades of purple or red reflect P deficiency in many plant species. Stunted or slow growth is another indication of poor crop nutrition.

In diagnosing nutritional disorders in wheat and other small grains, it is important to note the pattern and kind of symptoms on various plant parts and over the entire plant population in the field. Stunted, slow or uneven growth, loss of color, lack of tillering, low yields, shriveled grain, and premature senescence are common symptoms of nutrient disorders (Plate 65). All of these symptoms, however, may also arise from other causes such as temperature or moisture extremes and from stresses imposed by certain weeds, insects, and/or diseases.

In wheat, nutrient deficiency symptoms are most evident on immature plants and during periods of rapid vegetative growth. Demands for nutrients are highest during these periods and during phases of reproductive growth such as seed set and filling. Accurate diagnosis of nutritional disorders and competent fertility management require chemical analyses of the plant and/or soil in question in addition to an awareness of nutrient symptomology. Nutrient deficiencies identified are corrected by application of appropriate fertilizers. Nutrient excesses and toxicities are less easily corrected but are countered by leaching, extraction, degradation, or inactivation of the toxic entity.

Nutrients in Relation to Disease

Nutrients best support plant growth and best serve the interests of small-grain producers when their supply is optimal, neither deficient nor in excess. However, the optimal level of each nutrient for the most efficient and profitable production of small-grain cereals is sometimes difficult to define and challenging to maintain. Soil nutrient shortage or oversupply accentuates plant stress, abnormal growth, and predisposition to disease. Soilborne diseases of small grains such as the root and foot rots are often more visible and consequential in plants weakened by nutrient shortages. On the other hand, with an abundant supply of nutrients, small-grain plants may be more disease prone because they make dense vegetative growth and, in turn, provide a lush microclimate for microbial growth. In some instances of dryland culture, well-nourished plants grow more quickly, exhaust soil water supplies more rapidly, and become predisposed to disease because of drought stress.

In the paragraphs that follow, attention is focused on nutrient rather than disease disorders. However, reference is made to small-grain diseases that may be influenced by nutrient disorders where such relationships appear to be well documented.

Deficiency Symptoms

Nitrogen (N)

Nitrogen deficiency in small grains is first expressed through yellowing and then as stunted growth (Plate 65). Because N is readily translocated to metabolically active tissues, chlorosis begins on older tissues such as lower leaves (Plate 66). In older tissues, cell growth and division are slowed as is protein synthesis. As a consequence, starch, sugars, and amino compounds increase in cell sap. These materials can favor microbial growth and thereby predispose N-deficient plants to disease.

Nitrogen is an important element in proteins, chlorophyll, and various coenzymes. It is the nutrient most frequently deficient in plants. Despite the abundance of N in air, N is taken up primarily from soil and utilized by small-grain cereals as nitrate (NO_3^-) and to a lesser extent as ammonium (NH_4^+). Organic N is unavailable until converted to inorganic forms (mineralized) by microbial activity. Thus, N may be deficient in cool, wet, dry, or acid soils where microbial activity is limited.

Small grains and grasses, generally, are sensitive to insufficient N and responsive to supplemental N. Excess N promotes lush, rank, and prolonged vegetative growth. This circumstance can accentuate lodging, frost injury, and disease. In small grains, the demand for N is greatest during periods of rapid growth (jointing through heading) and declines toward maturity. Much of the N utilized for vegetative growth eventually is translocated to reproductive structures, especially to the developing seed. Nitrogen, especially as NO_3^-, is mobile in soil and, therein, is subject to depletion through leaching and denitrification (conversion of NO_3^- to gaseous N in the absence of oxygen). Conversely, NO_3^- may be increased through mineralization.

Phosphorus (P)

Wheat and small grains deficient in P are stressed and predisposed to diseases such as Pythium root rot. Unlike N-deficient plants, P-deficient plants maintain their green color and may be darker green than plants with sufficient P. However, they also are slow growing and late to mature. Leaf tips die back when shortages are severe (Plate 67), and the foliage of some cultivars may display shades of purple or red. Older tissues, such as in older leaves, are first to display P deficiency symptoms since the nutrient is translocated to metabolically active sites.

Cold wet soils are relatively inactive biologically, and P release from organic matter is slow or nonexistent.

Phosphorus applications are most successful in close proximity to roots. Band applications of P either with the seed or preplant (Plate 68) are more immediately available than broadcast applications. Excess P should be avoided, because absorption of Cu, Zn, and other nutrients may be adversely affected. On the other hand, a single heavy application of P can spur positive yield responses for several years.

Potassium (K)

Potassium is likely to be deficient in sandy, coarse-textured, intensively cropped soils and in soils that are highly weathered or leached. Potassium in soil is more mobile than P, less mobile than NO_3^-, and similar in mobility to NH_4^+. Like P, K is best utilized by plants when placed directly in the root zone, particularly at lower rates of application.

Potassium, like N, is translocated to metabolically active tissues, so shortages first appear in older tissues such as in lower leaves. Potassium chlorosis is at first uniform on older plant parts. Leaves, however, may eventually become streaked with yellow or appear scorched, bronzed, or blighted along their edges (Plate 69). Some cultivars, especially of barley, develop excessive numbers of tillers when K is deficient. Potassium-deficient small grains have weak straw and are prone to lodge. Diseases such as powdery mildew are accentuated on K-deficient plants.

Potassium deficiency is more apt to occur where straw and seed are both harvested from small-grain fields. Because the nutrient is only weakly translocated to seed, most K is retained in vegetative plant parts. Harvested grain normally contains approximately 0.5% K by weight.

Calcium (Ca)

Calcium deficiency has not been reported in small grains under field conditions. Induced Ca deficiency under experimental conditions causes graying, bleaching, rolling, and wilting of leaves (compare to Cu deficiency). Root growth especially is interrupted, and root tips may die. Root hairs, if produced, may not function either in nutrient or water uptake. Yield is markedly reduced even when foliar symptoms are inobvious.

Calcium forms an integral part of the structural tissues in leaves and is also a constituent of the middle lamella. Ca promotes ion uptake and the formation of root mitochondria. Ca also plays a role in mitosis, osmotic regulation, and membrane permeability.

Magnesium (Mg)

Magnesium shortages are extremely rare in small grains. Where they occur, they are associated with acid soils. A condition known as grass tetany in animals is an expression of deficient Mg in feed and forage. More often, therefore, Mg shortages are of greater concern to livestock producers than to small-grain producers.

Magnesium is an integral part of chlorophyll. When Mg is held deficient in controlled environments, plants lose color and become stunted. Just as with N deficiency, the yellowing occurs first on older leaves, because Mg is readily translocated to actively growing tissues and to developing seed.

Magnesium deficiencies in soil and plants can be corrected by applications of magnesium sulfate or potassium-magnesium sulfate. Foliar sprays do not appreciably increase Mg levels in foliage or seed. Dolomitic limestone is an inexpensive source of Mg that is especially used in acid soils. The limestone must be weathered and chemically degraded in soil to solubilize the Mg and make it available to plants.

Sulfur (S)

Symptoms of S deficiency in wheat and other small grains are very similar to those of N deficiency (Plates 65 and 66). Sulfur is typically deficient in mineral soils that are well drained, coarse textured, and low in organic matter.

Wheat and other small-grain plants that are deficient in S yield poorly and appear brightly chlorotic, yellow green, and stunted. Sulfur deficiency is typically first expressed in older tissues but eventually causes young leaves to become prominently yellow. Sulfur-deficient flour has reduced milling and nutrient qualities.

Sulfur deficiencies are indicated but not readily verified by soil tests. Tissue analyses often are necessary to confirm S deficiency and distinguish it from N deficiency. Sulfur deficiencies are corrected by applications of sulfate fertilizers or elemental S. Elemental S, however, is not water soluble and is largely unavailable to plants until converted to oxidized forms by microbial activity. Small grains require significantly less S than most other crops grown in rotation with them. Wheat, for example, requires about 1 lb of S/acre (1.12 kg/ha) for every 10-bu (27.2-kg) yield increment. Wheat seed at harvest contains about 0.1% S by weight. Usually 10–15 lb of S/acre (11–17 kg/ha) is sufficient for small-grain production.

Sulfur can be found in certain amino acids, coenzymes, and all proteins. It is directly involved in electron transport via sulfhydryl groups.

Iron (Fe)

Iron plays an important role in respiratory and photosynthetic reactions. Iron deficiency occurs especially in calcareous, alkaline soils (pH higher than 8.0). In these circumstances, plants can become variously chlorotic. When the resulting chlorosis develops within interveinal tissues, it produces distinctive yellow stripes on the leaves of small grains (Plate 70). Younger leaves and tissues are first to show such symptoms.

Because Fe readily forms complexes with other metallic cations, its availability and utilization by plants are influenced by the availability of elements such as Cu, Mn, Al, and P.

Soils usually contain large amounts of Fe; however, a relatively small amount is available for plant use. Iron becomes increasingly unavailable as soil pH increases or as imbalances or excesses of other nutrients occur. Additions of 1.8 lb/acre (2 kg/ha) of chelated Fe or 8 lb/acre (9 kg/ha) of ferrous sulfate ($FeSO_4$) will relieve most symptoms of Fe deficiency. Under high-pH soil conditions, a foliar application is often a more practical way to supply Fe to the plant and prevent soil tie-up.

Zinc (Zn)

Zinc deficiencies in small grains are rare. Where they occur, they usually are associated with sandy, calcareous, eroded soils low in organic matter and high in P. Zinc, like other nutrient elements, is less available in cool wet environments.

Plants deficient in Zn are stunted and produce few tillers. Leaves become chlorotic, especially midway between the margin and midrib, yielding a modest stripe (Plate 71) similar to that caused by insufficient Fe. Severe Zn deficiencies cause leaves to turn gray-white and die. Zinc plays a prominent role in auxin activity and, thereby, in the elongation of internodes.

Soils deficient in Zn are readily diagnosed by soil analysis. In many areas where Zn levels are below 0.5 ppm in soil, foliar or soil applications of either zinc sulfate or chelated Zn can restore green color and augment yield. A positive Zn response is most likely to occur where soils are high in P.

Manganese (Mn)

Manganese is most likely to be deficient or unavailable in highly organic soils and in high-pH, calcareous or sandy soils. Shortages of Mn in small grains are rare but are accentuated by cool, dry weather.

Manganese-deficient plants are chlorotic and slow to mature. Among small grains, oats appear to be most susceptible to Mn deficiency. Manganese-deficient wheat and oats develop small, roughly circular, gray-white spots on older leaves (Plate 72). Chlorotic gray-white specks and streaks may eventually appear and coalesce on younger tissues. Leaves may kink or droop at the base of the blade or wherever the spotting is intense.

Manganese affects both photosynthesis and respiration. Magnesium can sometimes function for Mn in interactions with Fe.

Copper (Cu)

Organic soils and soils high in P are apt to be low in Cu. Although required only in minute amounts, Cu is reported to be yield limiting in portions of the United States, Australia, and Western Europe. Copper is closely linked to Ca translocation and induces symptoms arising in part, if not primarily, from Ca deficiency. It also is closely linked to respiratory enzymes and oxidation-reduction reactions.

Copper-deficient plants are light green, with dry and twisted leaf tips (Plate 73). Chlorosis and bleaching are followed by death and further curling and twisting of leaf tips and margins. This "white tip" disorder is linked to Cu-deficient peat soils. Delayed maturity, shriveled grain, and distorted, twisted, incompletely emerged heads and awns may result. Roots of affected plants are stunted, excessively branched, and rosetted. Darkened nodes and areas on the neck may be part of the syndrome.

In some areas, soils with less than 3 ppm of Cu (determined by diethylenetriaminepentaacetic acid extraction) are considered Cu deficient. Desirable Cu levels for small-grain production usually exceed 5 ppm (5 mg/kg). In tissue analyses, mature leaves should contain more than 6 ppm and up to 20 ppm total Cu (6–20 mg/kg) on a dry weight basis. Harvested wheat contains about 12 ppm Cu by weight. This harvested Cu is most commonly replaced by additions of copper sulfate ($CuSO \cdot 5H_2O$) and Cu chelates. Formulations of Cu applied to foliage must be water soluble to be effective.

Boron (B)

Although B is deficient in many leached and acid soils, its effects on and requirement by cereals are minimal compared with those of other crops. Calcium appears to negatively influence the availability and activity of B. Grasses generally are not affected by low B levels. Where deficiencies have been induced, the leaves of small grains first appear bleached and later may become rolled, trapped along the culm or otherwise distorted. Heads and awns are similarly affected, especially in leached, acid soils (Plate 74). Flowering is impaired and anther production is interrupted, resulting in poor seed set and yield.

Boron accumulates especially in reproductive tissues. Flower parts are reservoirs for B and are the first tissues to be distorted and to malfunction when B is deficient. B appears to be required for germination of pollen grains and fertilization.

Chlorine (Cl)

Evidence that Cl is deficient in wheat or in other small grains under field conditions is emerging, but Cl deficiency symptoms remain poorly defined. Chlorine appears to be essential for small grains held in controlled environments; however, in the field, responses to Cl tend to be marginal and inconsistent. In some field instances, Cl applied to soil or foliage augments the yield of selected small-grain cultivars, but the mechanism for this response is unknown and apparently is dependent on environment, disease, and/or cultivar. Chloride applications can accelerate the rate of spike development in spring wheat and increase kernel weight, kernel volume, and test weight in wheat and barley.

Plants utilize Cl as the chloride ion (Cl^-). High levels of Cl can reduce leaf spotting and the impact of diseases such as take-all root rot, stripe rust, leaf rust, tan spot,

and Septoria leaf spot. Some of the yield responses previously attributed to K after KCl applications may in part have resulted from disease suppression by Cl. Chloride supplements are most beneficial in leached mineral soils and where salts and fertilizers containing Cl^- have not been previously applied. Chloride is highly water soluble and leachable.

To limit the incidence of take-all in the Pacific Northwest, applications of 36 lb of Cl/acre (40 kg/ha) are recommended for banding with seed or 70–125 lb of Cl/acre (80–140 kg/ha) are recommended for broadcast in early spring. Since the ammonium (NH_4^+) form of N also restricts take-all, the use of NH_4Cl fertilizer should be especially effective as a nutrient supplement and disease suppressant.

Molybdenum (Mo)

The requirement for Mo in small grains is smaller than for any other nutrient. Diagnostic symptoms specific for Mo deficiency are not well defined. Molybdenum is associated with N metabolism and therefore influences protein synthesis. When Mo is deficient, nitrates (NO_3^-) and amines (NH_2^-) accumulate in the tissues.

Most cereals and grasses are insensitive to Mo except at very low or high concentrations. If symptoms of a Mo disorder develop, they usually are in response to Mo excesses. Excess Mo complexes with anthocyanin to form blue granules in the cells of mottled older leaves.

Toxicity Symptoms

Aluminum (Al)

Aluminum is not an essential plant nutrient; however, Al may be growth-limiting because of its toxicity. Aluminum levels are highest in weathered, acid soils and may exceed the usual levels that most small grains can tolerate. While Al serves no nutritional purpose, it is still taken up and accumulated in many plant species. Aluminum has been implicated in reduced root and top growth, shallow root systems, and reduced Ca transport. Applications of lime, P, or chelating agents sometimes relieve Al toxicity.

Manganese

In acid soils (pH lower than 5), Mn solubility may increase to toxic levels. Excess Mn may provoke an Fe and chlorophyll deficiency. Symptoms of Mn toxicity and deficiency are similar, with leaves showing marginal chlorosis, curling, and necrotic spots.

Macronutrient Management

In soil and in small grains, most nutrient elements are chemically and physically interrelated so that adjustments to one may affect the requirement for or availability of others. In addition, the availability of all plant nutrients is influenced by soil moisture, temperature, and pH (Plate 75).

Because different crops utilize different amounts of nutrient elements, cropping frequency and sequence have an important effect on the supply and balance of soil nutrients. Wheat grown in rotation with corn or sorghum, for example, is likely to be more nutrient deficient than wheat grown in previously fallow soil or following a legume crop. Fortunately, levels of C, H, and O, over which we have little if any control, are rarely deficient.

Normally an annual soil test for N and for other important nutrient elements is justified. Soil samples should be collected during the period of most rapid plant growth. The samples should also extend to the rooting depth of the cereal crop (at least 24 in. [60 cm]) to detect residual N and other nutrients in the root zone. In addition to measuring inorganic N (NH_4^+, NO_3^-), it is wise to quantify any N that may be present during the growing season from organic sources (mineralization). In addition, any nutrient supplement should be balanced with available moisture, which, in most dryland production areas, is the primary yield determinant. If moisture is deficient, all or part of any nutrient supplement may go unused.

Nitrogen, for example, can be supplemented from sources such as urea, anhydrous ammonia, ammonium sulfate, ammonium nitrate, and aqueous ammonia. The quantity of such supplements should not only be consistent with available water but also with other crop inputs and with the yield potential of the small-grain cultivar in question. Using an estimate of potential yield, the total N requirement for a small-grain crop is the product of anticipated yield times the N requirement per unit of grain. The amount of N to be supplemented is the total N required minus the N available (mineralizable N plus soil test N).

Research has shown that between 2.5 and 3.8 lb/bu (5 and 7 kg/100 kg) of available N is needed for wheat production. Adequate N should be available, especially at tillering, when a large part of the yield potential is determined. Nitrogen applications at later growth stages tend to increase grain protein more than grain yield. Usually N applied to foliage at the onset of heading is readily translocated to developing seed. Both the quantity and quality of protein in wheat seed are related to N content and are of great importance to its milling and baking characteristics.

Because the effects of S can be masked by N supplementation, an N/S ratio of approximately 5:1 in soils is regarded as optimal. Sulfur is usually applied as ammonium sulfate, ammonium thiosulfate, as elemental S, or as a sulfate salt of K, Mg, or Ca.

Sulfur is taken up from soil primarily as SO_4^{2-} and in small amounts from the air as SO_2. Like N, SO_4^{2-} also is mobile in soil and plant tissues and is equally effective in banded or broadcast applications. It is readily soluble and mobile in water.

Phosphorus, unlike N, is relatively immobile in soil. Phosphorus is taken up from the root zone primarily as orthophosphate ($H_2PO_4^-$ and HPO_4^{2-}). Soils at pH above 8.0 or below 5.5 or soils low in organic matter and/or high in lime are likely to be deficient in P. Phosphorus does not move appreciably in soil, so it must be locally applied in the root zone if it is to be available to the plant. Once absorbed, it plays a major role in energy transfer associated with cell division, photosynthesis, and respiration.

Just as with N, a soil test is important for prescribing P fertilization. Maintaining soil P in the range of 10–20 ppm should be sufficient and cost effective for most crops. Mono- and diammonium phosphates, ammonium polyphosphate, superphosphate, phosphoric acid, and urea phosphates are common sources of P. Soil samples should be taken to a depth of at least 6 in. (15 cm), and taking numerous samples will avoid erratic results sometimes associated with previous banded applications of P.

Phosphorus use is inefficient in that only a fraction of that applied is utilized by the current crop. The remainder, however, should be largely available for subsequent crops. Harvested grain contains from 0.3 to 0.5% P by weight.

As with P, banding or placement with seed is an efficient means of applying K. The major and most economical sources of K are potassium chloride (KCl), potassium sulfate (K_2SO_4), and potassium-magnesium sulfate ($K_2SO_4 \cdot 2 MgSO_4$). The Cl in potassium chloride has some benefit in disease suppression, so yield responses following applications of potassium chloride may not entirely be attributable to K supplementation.

When soil K is higher than 150 ppm, wheat and other small grains typically show no response to supplemental K. Usual production situations benefit from an application of between 30 and 140 lb of K/acre (35 and 160 kg/ha) if soil test values are less than 150 ppm of soil K.

Selected References

1. Alloway, G. J., and Tills, A. R. 1984. Copper deficiency in world crops. Outlook Agric. 13:32-42.
2. Altman, D. W., McCuistion, W. L., and Kronstad, W. E. 1983. Grain protein percentage, kernel hardness, and grain yield of winter wheat with foliar applied urea. Agron. J. 75:87-91.
3. Barnes, J. S., and Cox, F. R. 1973. Effects of copper sources on wheat and soybeans grown on organic soils. Agron. J. 65:705-708.
4. Baskin, C. C., and Anderson, K. L. 1983. Wheat fertilization. Miss. State Univ. Agron. Dep. Inf. Sheet 1217.
5. Beaton, J. D., and Soper, R. J. 1986. Plant response to sulfur in western Canada. Pages 375-399 in: Sulfur in Agriculture. Monogr. 27. American Society of Agronomy,

Madison, WI.

6. Boyde, D. A., Yuen, L. T. K., and Needham, P. 1976. Nitrogen requirement of cereals. I. Response curves. J. Agric. Sci. 87:149-162.

7. Brown, J. C., Ambler, J. E., Chaney, R. L., and Foy, C. D. 1972. Differential responses of plant genotypes to micronutrients. Pages 389-418 in: Micronutrients in Agriculture. J. J. Mortvedt, P. M. Giordano, and W. L. Lindsey, eds. Soil Science Society of America, Madison, WI.

8. Buchholz, D. D. 1983. Soil Test Interpretations and Recommendations Handbook. Department of Agronomy, University of Missouri, Columbia.

9. Chapman, H. D. 1966. Diagnostic Criteria for Plants and Soils. Division of Agricultural Sciences, University of California, Riverside.

10. Chaudry, F. M., and Loneragan, J. F. 1970. Effects of nitrogen, copper and zinc fertilizers on the copper and zinc nutrition of wheat plants. Aust. J. Agric. Res. 21:865-879.

11. Christensen, N. W., and Brett, M. 1985. Chloride and liming effects on soil nitrogen form and take-all of wheat. Agron. J. 77:157-163.

12. Christensen, N. W., and Meints, V. W. 1982. Evaluating N fertilizer sources and timing for winter wheat. Agron. J. 74:840-844.

13. Cope, J. T., Jr. 1981. Effects of 50 years of fertilization with phosphorus and potassium on soil test levels and yields at six locations. Soil Sci. Soc. Am. J. 45:342-347.

14. Engel, R. E., Woodard, H., and Sanders, J. L. 1992. A summary of chloride research in the Great Plains. Pages 232-241 in: Proc. Conf. Great Plains Soil Fertility.

15. Engelhard, A. W., ed. 1989. Soilborne Plant Pathogens: Management of Diseases with Macro- and Microelements. American Phytopathological Society, St. Paul, MN.

16. Fales, S. L., and Ohki, K. 1982. Manganese deficiency and toxicity in wheat: Influence on growth and forage quality of herbage. Agron. J. 74:1070-1073.

17. Florell, C. 1957. Calcium, mitochondria and anion uptake. Physiol. Plant. 10:781-790.

18. Fleming, A. L. 1983. Ammonium uptake of wheat varieties differing in aluminum tolerance. Agron. J. 75:726-730.

19. Foy, C. D., Fleming, A. L., and Schwartz, J. W. 1973. Opposite aluminum and manganese tolerances in two wheat cultivars. Agron. J. 65:123-126.

20. Gardner, E. H., Jackson, T. L., Wilcox, B. G., Fitch, L., Johnson, M., and Pumphrey, V. 1983. Fertilizer guide: Irrigated wheat. Oreg. State Univ. Ext. Serv. FG 40.

21. Giordano, P. M., Noggle, J. C., and Mortvedt, J. J. 1974. Zinc uptake by rice as affected by metabolic inhibitors and competing cations. Plant Soil 41:637-646.

22. Goos, R. J., Westfall, D. G., Ludwick, A. E., and Goris, J. E. 1982. Grain protein content as an indicator of N sufficiency for winter wheat. Agron. J. 74:130-133.

23. Halvorson, A. D., and Black, A. L. 1985. Long-term dryland crop responses to residual phosphorus fertilizer. Soil Sci. Soc. Am. J. 49:928-933.

24. Hanson, R. G., Maledy, S. R., Hoette, R. D., Watchinski, N. W., and Hentes, C. E. 1984. Nitrogen source, time of application, and nitrification inhibitor effect on yield of winter wheat. J. Fert. Issues 1:54-61.

25. Huber, D. M. 1981. The use of fertilizers and organic amendments in the control of plant diseases. Pages 357-394 in: Handbook of Pest Management in Agriculture. D. Pimentel, ed. CRC Press, Boca Raton, FL.

26. Jordan, H. V., and Ensminger, L. E. 1958. The role of sulphur in soil fertility. Adv. Agron. 10:407-434.

27. Krantz, B. A., and Melsted, S. W. 1964. Nutrient deficiencies in corn, sorghum and small grains. Pages 25-58 in: Hunger Signs in Crops. H. B. Sprague, ed. David McKay, New York.

28. Lamb, C. A. 1967. The nutrient elements. Pages 207-217 in: Wheat and Wheat Improvement. K. S. Quisenberry and L. P. Reitz, eds. Monogr. 13. American Society of Agronomy, Madison, WI.

29. Leikam, D. F., Murphy, L. S., Kissel, D. E., Whitney, D. A., and Moser, H. C. 1983. Effects of nitrogen and phosphorus application method and nitrogen source on winter wheat yield and leaf tissue phosphorus. Soil Sci. Soc. Am. J. 47:530-535.

30. Makarim, A. K., and Cox, F. R. 1983. Evaluation of the need for copper with several soil extractants. Agron. J. 75:493-496.

31. McDole, R. E. and Mahler, R. L. 1985. Northern Idaho fertilizer guide: Winter wheat. Univ. Idaho Coop. Ext. Serv. CIS 453.

32. McDole, R. E. and Mahler, R. L. 1985. Northern Idaho fertilizer guide: Feed barley. Univ. Idaho Coop. Ext. Serv. CIS 758.

33. Ohki, K. 1984. Manganese deficiency and toxicity effects on growth, development and nutrient composition in wheat. Agron. J. 76:213-218.

34. Papastylianou, I., Graham, R. D., and Puckridge, D. W. 1982. The diagnosis of nitrogen deficiency in wheat by means of a critical nitrate concentration in stem bases. Commun. Soil Sci. Plant Anal. 13:473-485.

35. Peterson, G. A., Sander, D. H., Grabouski, P. H., and Hooker, M. L. 1982. Efficient use of phosphorus on winter wheat. Better Crops Plant Food 62:3-5.

36. Rasmussen, P. E., and Allmaras, R. R. 1985. Sulfur fertilization effects on winter wheat yield and extractable sulfur in semi-arid soils. Agron. J. 78:421-425.

37. Rasmussen, P. E., and Kresge, P. O. 1986. Plant response to sulfur in the Western U.S. Pages 357-374 in: Sulfur in Agriculture. Monogr. 27. American Society of Agronomy, Madison, WI.

38. Robinson, J. B. D. 1984. Diagnosis of Mineral Disorders in Plants, vols. 1 and 2. Chemical, New York.

39. Schaff, B. E., and Skogley, E. O. 1982. Soil profile and site characteristics related to winter wheat response to potassium fertilizers. Soil Sci. Soc. Am. J. 46:1207-1211.

40. Snowball, K., and Robson, A. K. 1991. Nutrient Deficiencies and Toxicities in Wheat: A Guide for Field Identification. Centro Internacional de Mejoramiento de Maiz y Trigo (CIMMYT), Mexico City.

41. Stanford, G. 1982. Assessment of soil nitrogen availability. Nitrogen in Agricultural Soils. F. J. Stevenson, ed. Agronomy 22:561-720.

42. Steward, F. C. 1963. Plant Physiology. Vol 3, Inorganic Nutrition of Plants. Academic, New York.

43. Unruh, L. G., and Whitney, D. A. 1984. Varying aluminum tolerance in wheat grown in south central Kansas. 1984 Kansas Fert. Rep. 462:14-16.

44. Wallace, S. U., and Anderson, I. C. 1984. Aluminum

toxicity and DNA synthesis in wheat roots. Agron. J. 76:5-8.

45. Walsh, L. M., and Beaton, J. D. 1973. Soil Testing and Plant Analysis. Soil Science Society of America, Madison, WI.

46. White, R. P. 1970. Effects of lime upon soil and plant manganese levels in acid soil. Soil Sci. Soc. Am. J. 34:624-629.

47. Wrigley, C. W., du Cros, D. L., Moss, H. J., Randall, P. J., and Fullerton, J. G. 1984. Effect of sulfur deficiency on wheat quality. Sulphur Agric. 8:2-7.

48. Younts, S. E. 1964. Response of wheat to rates, dates of application and sources of copper and to other micronutrients. Agron. J. 56:266-269.

Part II

Sugar and Oilseed Crops

Chapter 6

Gary J. Gascho
Department of Crop and Soil Sciences
University of Georgia, Tifton

David L. Anderson
Everglades Research and Extension Center
University of Florida, Belle Glade

John E. Bowen
Department of Plant Molecular Physiology
University of Hawaii, Honolulu

Sugarcane

Most modern sugarcane cultivars are interspecific hybrids of three species of the genus *Saccharum* (*S. officinarum, S. spontaneum,* and *S. robustum*). Approximately 27 million acres (11 million hectares) are devoted to the cultivation of this crop throughout the tropical and subtropical countries of the world. Sugarcane is traditionally grown in monoculture for the production of crystalline sucrose. Ethanol, used principally as a gasoline additive, has become an important product in some areas. Numerous by-products are also important.

Photosynthesis in sugarcane is via the C_4 pathway, a factor contributing to the extraordinary efficiency of this plant in converting solar energy into biomass. Since sugarcane has a long growing season (ranging from 8 to 36 months worldwide) and an extensive root system, nutrient uptake is highly efficient. Nutrient uptake and utilization are normally slow during the 1–12 weeks after planting or ratooning (regrowth after harvest). During the grand growth period (12–32 weeks after planting), nutrient uptake and plant development are rapid. The later maturation period is characterized by low nutrient uptake and low biomass accumulation; however, during this period, sugar assimilates accumulate. This is referred to as the ripening period, which elevates sugar yields. Growth periods are variable and depend on regional climatic conditions. For example, in Louisiana, sugarcane is grown from 7 to 10 months. Total sugarcane production in Louisiana averages 23 tons/acre (52 t/ha). In contrast, sugarcane grown in Hawaii is grown from 15 to 24 months, with some sugarcane at high elevations grown for 36 months or longer. Total sugarcane production in Hawaii averages 96 tons/acre (215 t/ha). With such variation on a single-harvest basis, nutrient needs and removal data per unit of land area are not meaningful. Needs are best portrayed on a yield goal basis considering cultivar, soil, nutrient reserves, and climatic conditions.

Barnes (2) reported removal of 1.5–1.8 lb of N/ton of cane (0.7–0.9 kg/t), 0.44–0.52 lb of P/ton of cane (0.22–0.26 kg/t), and 2.49 lb of K/ton of cane (1.28 kg/t). In contrast, nutrient assimilation values in Hawaii (7) per ton of cane were were 0.96 lb of N (0.48 kg/t), 0.19–0.66 lb of P (0.09–0.33 kg/t), and 1.5 lb of K (0.75 kg/t). These differences may be accounted for by the amounts of trash (leaves and roots) transported to the mill, age of sugarcane at harvest, and nutrient uptake differences among cultivars. Low-yielding cane often accumulates nutrients at rates similar to high-yielding cane, and therefore the concentrations of mobile and immobile nutrients are higher than in rapidly accumulating biomass. Sugarcane is different from other crops in several ways: the length of crop growth varies around the world; there is a large genetic pool of cultivars accounting for differences in nutrient uptake, disease, yield, and growth; and only the stalk is used in processing. Because of these differences, uptake information may not always be meaningful even within one geographic region.

Fertilization needs are affected by the management of residues and wastes generated in processing. The crop is most often burned in the field before cutting. Although N is largely lost, ash remaining in the field supplies soluble K and P. After a number of crops, burning will decrease the soil organic reserves added in the sugarcane monoculture.

Waste materials derived from milling cane, including bagasse, filter cake, waste waters, and distillery slops, can be important sources of nutrients for subsequent crops. Approximately 25% of the milled stalks is bagasse, a cellulitic fiber left after juice extraction. Bagasse has a low nutrient amendment value with variable analyses. Bagasse concentrations determined in Florida (1) are approximately 0.32% N, 0.09% P, and 0.44% K. Mills burn bagasse to produce energy for their operations. Excess bagasse is a source of industrial chemicals and is also used in the manufacture of fibrous materials. Filter cake, another by-product, is generated at the rate of 90–100 lb (40–45 kg) for each ton (0.91 t) of cane processed. The respective N, P, and K concentrations of filter cake are approximately 1.0, 0.5, and 0.8%. In areas of the world where large quantities of ethanol are distilled from sugarcane juice, distillery slops, often called "dunder" or "venase," is a by-product. Fourteen volumes of slops are produced for each volume of ethanol. This material has a dry ash value of 31% K and is an important source of K applied to fields in Brazil. All of these by-products have a low nutrient value, and raw materials would have to be applied at very high rates to be of equivalent value as a fertilizer. Although some by-products may improve soil physical properties, amendment use has little effect on the nutrition of cane.

Deficiency Symptoms

The following description of deficiency symptoms is based on work done by Gascho and Taha (12) and Anderson and Bowen (1).

Nitrogen (N)

Older leaves die back. Leaf blades of N-deficient plants turn uniformly light green to yellow (Plate 76). Stalks become short and slender, and the vegetative growth rate is reduced. Tips and margins of older leaves become necrotic prematurely.

Phosphorus (P)

Since the effect of P deficiency is retardation of plant growth, distinct foliar symptoms are not always observed. Phosphorus is mobile in the plant; therefore, deficiency first occurs in older tisues. Older leaves die back. Leaf blades become dark green to blue-green. Red or purple colors often appear, particularly at tips and margins exposed to direct sunlight. The leaves are slender. Older leaves turn yellow and eventually die back from tips and along the margins (Plate 77). Stalks are short and slender. Internodal length is greatly reduced toward the top of the plant. Plant vigor and tillering are reduced. Millable stalk number and sugarcane tonnage are reduced, particularly in ratoons. Deficiency is much more common in ratoon fields, and its intensity tends to increase with age of the planting.

Potassium (K)

Necrotic lesions appear between veins along margins and tips of older leaves. Older leaves may be entirely brown or "fired" (Plate 78). Red discoloration may appear on the upper surface of leaf midribs (Plate 79). Young leaves usually remain dark green. The leaf spindle will distort, producing a characteristic "bunched top" or "fan" appearance. Stalks become slender.

Calcium (Ca)

Both young and old leaves are affected by Ca deficiency. Old leaves may have a "rusty" appearance similar to that seen in Mg deficiency and may die prematurely. When young leaves are affected, apical meristems die back, and immature leaves are distorted and necrotic. Young leaves hook or turn under, and spindles are often necrotic at the tip and along margins when the deficiency is acute. Minute chlorotic lesions with necrotic centers form on young leaves and later turn reddish brown. The rind may become soft, and stalks will be slender and taper rapidly toward the growing point.

Magnesium (Mg)

Mottling or chlorosis begins at the tip and along margins of older leaf blades of Mg-deficient plants. Red necrotic lesions appear, resulting in a rusty appearance (Plate 80). The rind may exhibit internal browning similar to Ca deficiency.

Sulfur (S)

As with N deficiency, S deficiency causes leaves to become uniformly chlorotic (Plate 81). Leaves become narrower and shorter than normal. In contrast to N deficiency, the youngest leaves are most affected and more chlorotic than the older leaves, because S is not mobile in the plant. Stalks will be slender, resulting in reduced tonnage of millable cane after a prolonged deficiency. In subtropical areas, S deficiency is often temporary during periods of suboptimal temperature.

Iron (Fe)

Immature leaf blades are distorted and necrotic. Interveinal chlorosis occurs from the tip to base of leaves (Plate 82). The entire plant may become chlorotic (Plate 83). Ratoon plants exhibit Fe deficiency when root development is slow. Plants normally outgrow symptoms, but with excessive root damage and low soil Fe conditions, the deficiency symptom may persist. Foliar sprays can be used to correct chlorotic leaf symptoms, but they may not increase yields.

Zinc (Zn)

Veinal chlorosis begins at the base of young leaf blades (Plate 84). Leaves are small and nonsymmetrical. Irregular horizontal "rippling" of leaf blades may occur. Necrosis may occur when Zn deficiency is severe, spreading from the tip to the base of the leaf blade. Plants have reduced tillering and ratooning abilities.

Manganese (Mn)

Immature leaf blades of Mn-deficient plants have varied degrees of chlorosis but are not wilted. Interveinal chlorosis occurs from the tip toward the middle of the leaf (Plate 85). Chlorotic stripes may bleach and fray in the wind.

Copper (Cu)

Green splotches appear on the leaf blades of Cu-deficient plants (Plate 86). Leaves eventually bleach and become papery thin and rolled when deficiency is severe. Stalks and meristems sometimes lack turgidity and will have a wilted "rubbery" appearance ("droopy-top" disease).

Boron (B)

Immature leaves are distorted and necrotic. On the distorted leaves, translucent lesions or "water sacks" form between veins (Plate 87). Leaf tips may be severely burned (Plate 88). Young plants are bunched, with many tillers. Leaves tend to be brittle, and spindle leaves may be chlorotic and later necrotic. This malady is sometimes referred to as "false pokkah boeng" disease because of the similarity of the spindle leaf symptom. This symptom can also be confused with herbicide damage.

Chlorine (Cl)

Young elongating leaves wilt reversibly, especially on hot, sunny days, but usually recover overnight. Roots are abnormally short with an increase in the number of lateral branches.

Molybdenum (Mo)

Older leaves die back. Leaf blades are uniformly light green to yellow. Stalks become short and slender, and vegetative growth rate is reduced. Short longitudinal chlorotic streaks begin on the apical one-third of the leaf. Older leaves dry prematurely from the middle toward the tip.

Silicon (Si)

Silicon deficiency produces minute circular white leaf spots (freckles), which are more severe on old leaves (Plate 89). Older leaves may senesce prematurely. The plant will exhibit poor tillering and ratooning characteristics.

Toxicity Symptoms

Aluminum (Al)

In plants affected by Al toxicity, few lateral roots form; those that do have abnormally thickened tips. Damage to roots resembles that caused by nematodes. Plants become highly susceptible to moisture stress and other nutritional maladies.

Boron

Chlorosis of tips and margins spreads on young leaf blades of plants with excess B. Chlorosis ultimately extends to older leaves. Chlorotic tissue quickly becomes necrotic.

Chlorine

Roots are abnormally short, but unlike deficiency of Cl, there is very little lateral root branching.

Sodium (Na)

High accumulations of Na in the plant adversely affect root and top growth. Leaf tips and margins will dry out and have a scorched appearance. Excessively high Na concentrations decrease the chlorophyll content of leaves, lowering the net photosynthesis per unit leaf area. Under these conditions, leaves may have a pale green to yellowish green appearance. High Na is associated with high Cl levels.

Sulfur Dioxide (SO$_2$)

Sulfur dioxide toxicity is characterized by mottled chlorotic streaks running the full length of leaf blades. The leaf tips and margins may become necrotic. These symptoms occur within 3 to 7 days after exposure to SO$_2$ gases. Older leaves are not usually affected. Toxicities are most commonly observed where there is volcanic activity.

Critical and Sufficient Concentrations and Nutrient Balance

Nutrient Concentration

Although the goals of tissue analysis may be similar, choices of the diagnostic tissues, sampling times, and interpretation differ among methods. Worldwide, the choice of tissues used in interpretation include leaf blades, leaf sheaths, stalk internodes, and roots. Selection of leaf tissues also varies and includes tissues from the youngest leaf lacking a functional ligule or dewlap formation, mature leaf tissues of increasing age and position from the meristem, and collective tissues from different positions. These tissues have been often referred to as the "visible dewlap" leaves, whereby the top visible dewlap (TVD) leaf position is the first mature leaf blade from the apical meristem (top) having a functional ligule (dewlap). Nutrient analyses are made on these tissues, using whole leaf blades, leaf blades without midrib tissues, and punches and selected portions of the leaf blade. Where the system of crop logging (7) is used, collective tissues from elongating blades and sheaths from leaf positions 3 to 6 are used. It is obvious that the system of leaf sampling is not uniform around the world and that variation in reported results and interpretation are also different.

Ranges in adequacy, critical levels, and excessive levels for nutrient concentrations in TVD leaf blades with their midribs removed in Florida sugarcane are given in Table 6.1. Critical and sufficient levels in the TVD leaf blade vary to some extent in studies from a number of countries (10,11). In addition to some variations by geographic region, significant differences have been found (18) in

N, P, Ca, and Mg concentrations in TVD leaf blades resulting from within-day sampling time (Table 6.2). Concentrations decreased about 10% from 8 A.M. to 5 P.M. Florida cultivars and age of cane at sampling also significantly affected TVD leaf blade nutrient concentrations (18). Samples must be taken at a specific time during the day in order to reduce variability, preferably very early in the day. For routine crop monitoring, samples need to be restricted to the grand growth period, i.e., when sugar cane is growing at its maximum rate. Other sampling is recommended on tissues exhibiting nutritional abnormalities during the period of observation.

Plant analysis for assessment of sugarcane nutritional status is used extensively; however, the approach is not consistent. The tissue used and timing of sampling affect the analytical results and their interpretation. A system referred to as "crop logging" was developed in Hawaii (7). This system involves repeated samplings and analyses over the crop cycle. Data collection begins when the crop is 3 months old, and samples of leaf parts are taken every 5 weeks. Data collected include growth rate, light intensity, temperature, rainfall or irrigation, total sugars, fertilization and field culture records, harvest data, tissue moisture index, and tissue analyses. Data are plotted as a permanent record for a given field. This system is useful for making decisions about fertilization, irrigation, and harvest and schedules for crops exceeding 12 months of growth. Most sugarcane is not grown as intensively as is assumed by the crop log system, so plant tissue analysis is generally practiced on a reduced scale to monitor fertilization practices and diagnose nutritional maladies.

When six different plant parts were compared, significant differences were found in nutrient concentrations (18). Plant parts sampled included the TVD leaf blade with midrib, the TVD leaf blade without midrib, the middle of the TVD leaf blade, the middle of the elongating leaf blade, the elongating leaf sheath, and the whole aboveground plant. Concentrations in all tissues were correlated with concentrations in the aboveground whole plant, but concentrations and ratios in the TVD leaf blades were more highly correlated with sugarcane tonnage at harvest than were the concentrations in the sheath or the whole plant. Nutrient concentrations in the whole TVD leaf blade and the middle portion of the same blade were not statistically different. We concluded that the practice of subsampling the middle portion of the blade appears unnecessary. We also found that the highest precision (lowest coefficient of variation) was obtained using the TVD leaf blade. Therefore, since the TVD leaf blade is a well-defined and easily accessible tissue, we concluded that the TVD leaf with or without its midrib is the tissue of choice in Florida. Likewise, the TVD leaf blade has been selected for use in Mauritius (13), Jamaica (14), and South Africa (4).

Nutrient Balance

Nutrient interactions and balances may be as important as actual concentrations as evidenced by recent studies on the diagnosis and recommendation integrated system (DRIS) first proposed by Beaufils (3). Studies on the DRIS are of interest because the system moderates the influences of crop age at sampling, cultivar, and geography. The DRIS allows the comparison of all analyzed nutrients with each other and then ranks them from negative to positive. The element with the most negative ranking is then declared the most likely to be limiting, the second most negative ranking is declared to be the next most limiting nutrient, etc. The system is qualitative until one decides how negative a ranking must be to ensure that a deficiency or an imbalance actually exists. Details of the DRIS procedure cannot be presented here but are available elsewhere (3,19).

The applicability of DRIS to sugarcane has been the subject of several studies (4,5,8,9,15). Beaufils and Sumner (5) sampled leaf blades at 3, 6, 12, and 18 months and concluded that DRIS was useful over a wide range of cane age and that the critical nutrient level approach was not. In Hawaii, the DRIS provided slightly more

Table 6.1. Nutrient concentrations in TVD leaf blade laminas collected in Florida sugarcane during grand growth[a,b]

Nutrient	Range	Critical	Normal	Excessive
N, %	1.90–2.20	1.80	2.00–2.60	>3.20
P, %	0.19–0.22	0.19	0.22–0.30	>0.34
K, %	1.02–1.35	0.90	1.00–1.60	>2.20
Ca, %	0.16–0.20	0.20	0.20–0.45	>0.50
Mg, %	0.08–0.19	0.12	0.15–0.32	>0.35
Si, %	0.1–5	1.00	...	...
Fe, ppm	50–200	...	50–105	>105
Mn, ppm	25–400	...	12–100	>100
Zn, ppm	15–25	15	16–32	>40
Cu, ppm	<4	3	4–8	>9
B, ppm	0.4–34	4	...	>45

[a]Data from Gascho and Elwali (11) and Kidder and Gascho (16).
[b]TVD = top visible dewlap.

Table 6.2. Mean tissue nutrient concentrations in sugarcane TVD leaf blade laminas as a function of time of day[a,b]

Time of day	Plant tissue nutrients				
	N	P	K	Ca	Mg
8 A.M.	2.03	0.24	1.41	0.28	0.20
11 A.M.	1.97	0.25	1.40	0.27	0.19
2 P.M.	1.88	0.23	1.38	0.26	0.18
5 P.M.	1.80	0.22	1.40	0.25	0.18
Significance[c]	**	**	NS	**	**

[a]Source: Thein (17).
[b]TVD = top visible dewlap.
[c]The 1% level of significance of the linear regression is indicated by **. NS = not significant.

accurate diagnoses than the conventional crop log approach (15). Nutrient ratio norms developed separately in South Africa and Florida are similar despite large differences in soils, cultivars, climate, and other factors (Table 6.3).

Elwali and Gascho (9) developed norms for nine essential elements for sugarcane and compared the conventional soil test method used in Florida with foliar analysis interpreted both by critical nutrient level and DRIS approaches. We sampled eight fields in the early portion of grand growth and applied fertilizers according to each of the three methods. Fertilization according to the DRIS approach significantly improved nutrient balance and also increased cane and sugar yields in comparison to the other two methods. The DRIS appeared to be a good method for diagnosis of needed nutritional corrections in the current crop as well as for future crops.

Nutrient Interactions with Crop Management

Many conditions affect sugarcane and interact with nutrient balance. These conditions may cause symptoms of nutrient imbalance that are unrelated to a soil fertility requirement. Such conditions include overfertilization, drought, flood, high and low temperatures, physical injury to the plant by tillage equipment, nematodes, diseases, insects (i.e., *Ligyrus subtropicus, Diatraea saccharalis, Diaprepes abbreviatus*), herbicides, sodic soils, acid soils, and alkaline soils. Interactions of some nutrients with crop management and environmental conditions are discussed briefly.

Soil pH interacts strongly with nutrient availability. Sugarcane is grown on soils of pH 4–8 and is often judged as nonresponsive to pH (a misconception based on some other plants that are more sensitive). At low pH, Al and Mn can be toxic. The availability of P, Ca, and Mg are reduced at low pH. In the upper portion of the pH range, Fe, Zn, Mn, Cu, and B decrease in availability and can be deficient for optimum growth, whereas the availability of Mo increases. Soil pH also affects N loss and its management. Gaseous losses of surface-applied urea and other ammoniacal N fertilizers increase with soil pH because of urea hydrolysis and reaction with $CaCO_3$ to form precipitates and ammonium bicarbonates, which are readily disassociated, resulting in N lost as ammonia.

Soil texture, exchange capacity, and colloidal (organic matter and clay) contents are important factors in interactions with nutrients. In general, nutritional problems and the intensity of management required increase on sandy soils, which have low exchange capacities and low colloid contents. Nitrogen and B are particularly leachable on such soils. Nitrogen must be applied in split applications to avoid losses and achieve adequate uptake. Native organic soils are often deficient in Cu, but the deficiency can be corrected for a period of years by a single application of copper sulfate.

Aluminum toxicity occurs on some acidic soils when the pH is less than 5.2. Crops grown on soils high in Al minerals (i.e., bauxite), resulting in low pH and high Al saturation in soil solution, are particularly susceptible. Tolerance to Al toxicity is found among some cultivars, and some are now bred to tolerate high Al. Aluminum toxicity can accentuate Ca and P deficiencies by precipitating alumino-calcium-phosphate complexes. Because Al toxicity adversely affects root growth, the effective nutrient absorbing power of roots is reduced. This may result in deficiencies of other nutrients (e.g., Mg, Fe, Mn, etc.), reduction of drought tolerance, and increased susceptibility to insects and pathogenic diseases. Higher P fertilization, liming, and gypsum applications often alleviate Al toxicity.

Silicon is considered a functional rather than an essential element. Leaf freckling, caused by Si deficiency, may be increased by liming with $CaCO_3$ or fertilizing with phosphate. Leaf freckling appears to be cultivar specific and may not be expressed in some cultivars even when their Si concentrations are low. Large responses to the application of calcium silicates have been observed in Florida and Hawaii on soils low in soluble silicates. In Hawaii, a Mn/SiO_2 ratio above 70 indicates that an application of silicate should be considered.

Table 6.3. Sugarcane TVD leaf blade lamina norms from Florida and South Africa[a,b]

Nutrient ratio	Florida	South Africa
N/P	8.706	8.197
N/K	1.526	1.511
K/P	5.633	5.464
Ca/N	0.151	0.128
Ca/P	1.314	1.146
Ca/K	0.222	0.205
Ca/M	1.373	1.158
Mg/N	0.113	0.116
Mg/P	0.984	0.962
Mg/K	0.163	0.186

[a] Data from Beaufils and Sumner (4) and Elwali and Gascho (8).
[b] TVD = top visible dewlap.

Selected References

1. Anderson, D. L., and Bowen, J. E. 1990. Identification key to the malnutrition of sugarcane. In: Sugarcane Nutrition. Foundation for Agronomic Research and Potash and Phosphate Institute, Atlanta, GA.
2. Barnes, A. C. 1974. The Sugar Cane. John Wiley & Sons, New York.
3. Beaufils, E. R. 1973. Diagnosis and recommendation integrated system (DRIS). Univ. Natal Soil Sci. Bull. 1.
4. Beaufils, E. R., and Sumner, M. E. 1976. Application of the DRIS approach for calibrating soil and plant factors

in their effects on yield of sugarcane. Proc. S. Afr. Sugar Technol. Assoc. 50:118-124.

5. Beaufils, E. R., and Sumner, M. E. 1977. Effect of time of sampling on the diagnosis of the N, P, K, Ca, and Mg requirements of sugarcane by the DRIS method. Proc. S. Afr. Sugar Technol. Assoc. 51:131-137.

6. Bowen, J. E. 1977. Essentiality of chlorine for optimum growth of sugarcane. Proc. Congr. Int. Soc. Sugar Cane Technol. 14:1113-1120.

7. Clements, H. F. 1980. Crop Logging of Sugarcane—Principles and practices. University of Hawaii Press, Honolulu.

8. Elwali, A. M. O., and Gascho, G. J. 1983. Sugarcane response to P, K, and DRIS corrective treatments on Florida Histosols. Agron. J. 75:79-83.

9. Elwali, A. M. O., and Gascho, G. J. 1984. Soil testing, foliar analysis, and DRIS as guides for sugarcane fertilization. Agron. J. 76:466-470.

10. Evans, H. 1965. Tissue diagnostic analyses and their interpretation on sugarcane. Proc. Int. Soc. Sugar Cane Technol. 12:156-180.

11. Gascho, G. J., and Elwali, A. M. O. 1978. Tissue testing of Florida sugarcane. Univ. Fla. IFAS Belle Glade AREC Res. Rep. EV-1978-3.

12. Gascho, G. J., and Taha, F. A. 1972. Nutrient deficiency symptoms of sugarcane. Univ. Fla. Agric. Exp. Stn. Circ. S-221.

13. Halais, P. 1950. Foliar diagnosis, a new guide to fertilization of sugarcane in Mauritius. Proc. Int. Soc. Sugar Cane Technol. 7:218-232.

14. Innes, R. F. 1948. Plant analysis in relation to the nutrition of sugarcane in Jamaica. Proc. Meet. West Indies Sugar Technol. pp. 92-114.

15. Jones, C. A., and Bowen, J. E. 1981. Comparative DRIS and crop log diagnosis of sugarcane tissue analysis. Agron. J. 73:941-944.

16. Kidder, G., and Gascho, G. J. 1977. Silicate slag recommended for specified conditions in Florida sugarcane. Univ. Fla. Coop. Ext. Serv. Agric. Facts 65.

17. Thein, S. 1979. Comparison of six tissues for diagnosis of sugarcane mineral nutrient status. M.S. thesis. University of Florida, Gainesville.

18. Thein, S., and Gascho, G. J. 1980. Comparison of six tissues for diagnosis of sugarcane mineral nutrient status. Proc. Congr. Int. Soc. Sugar Cane Technol. 17:152-162.

19. Walworth, J. L., and Sumner, M. E. 1987. The diagnosis and recommendation integrated system (DRIS). Adv. Soil Sci. 6:149-188.

1. Progressive N deficiency in corn (normal leaf on right). Note yellowing down the midrib.

2. Nitrogen deficiency in corn, beginnning on older leaves and progressing upward.

3. Phosphorus deficiency in corn, with reddish purple tips and leaf margins.

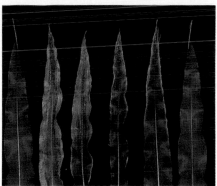

4. Progressive K deficiency in corn (normal leaves on outside), with yellowing and necrosis of leaf margins.

5. Potassium deficiency in corn, with yellowing and necrosis progressing from older to younger leaves.

6. Severe K deficiency in corn, with yellowing of lower leaves while upper leaves remain green.

7. Calcium deficiency in corn, with leaves failing to unfold.

8. Interveinal chlorosis caused by Mg deficiency in corn.

9. Sulfur deficiency in corn, with general yellowing of young leaves.

10. Iron deficiency in corn. Interveinal areas of young leaves are pale green to white.

11. Zinc-deficient corn, with interveinal light striping beginning at the base of younger leaves.

12. Manganese deficiency in corn, with young leaves becoming olive green and slightly streaked.

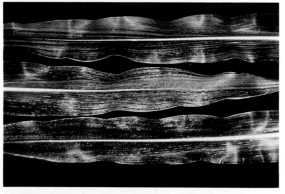

13. Boron-deficient corn leaves.

14. Corn leaves with necrotic edges from B toxicity.

15. Row skips due to ammonia injury in corn.

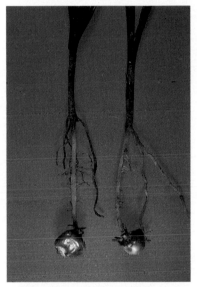

16. Ammonia injury to corn roots (left), causing poor or retarded growth.

17. Potassium deficiency in corn caused by soil compaction and restricted root growth.

18. Phosphorus deficiency symptoms in corn roots caused by herbicide injury.

19. Nitrogen deficiency in rice.

20. Phosphorus deficiency in rice.

21. Premature necrosis of rice leaves due to P deficiency.

22. Potassium deficiency in rice.

23. Chlorosis in upper leaves of S-deficient rice.

24. Response of rice growing on a muck soil to Fe applied in the row.

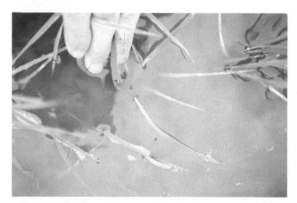

25. Zinc deficiency in seedling rice.

26. Zinc deficiency in rice caused by deep flooding.

27. Manganese deficiency in rice growing in a Florida Histosol.

28. Brown spot in rice induced by Si deficiency.

29. Salinity damage in rice.

30. Straighthead symptoms on rice panicle.

31. Straighthead in rice induced by residue from an arsenical herbicide.

32. Young field-grown sorghum plants with N deficiency symptoms on lower leaves.

33. Young field-grown sorghum plants with P deficiency. Note reddish color on lower leaves and on the stalk and sheath of leaves.

34. Sorghum plant with K deficiency on leaves. Lower leaves show more severe symptoms than upper leaves. Note lesions and necrosis on margins and leaf tips.

35. Close-up of sorghum leaves with excess P and with Mg, K, and Mn deficiencies (top to bottom). Note the difficulty in resolving color differences on plants with these nutrient disorders.

36. Calcium deficiency in top leaves of greenhouse-grown sorghum plants. Note serrated, torn, and slightly yellow-white leaves.

37. Young sorghum plant with Mg deficiency. Symptoms on lower leaves are most severe.

38. Young field-grown sorghum plant with S deficiency developing in emerging leaves.

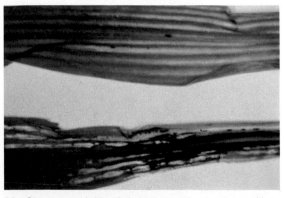

39. Close-up of Mn-deficient sorghum leaves. Note greater severity of symptoms on lower leaf and long brown streaks that later turn dark reddish brown.

40. Young field-grown sorghum plants with relatively severe Fe deficiency symptoms. Note long yellow streaking in uppermost leaves and leaves beginning to turn completely yellow.

41. Severe Fe deficiency on sorghum plants. Small plots were sprayed with Fe to overcome the disorder; note distinct differences between treated and untreated plants.

42. Young field-grown sorghum plant with severe Zn deficiency. Note uniform streaking and enhanced bleaching on the emerging leaf.

43. Young glasshouse-grown sorghum plants with 0, 8.8, and 22.2 mg of Cu per container. Note bending and drooping of leaves, especially tips. Calcium-deficient plants show similar symptoms if deficiency is not severe.

44. Young glasshouse-grown sorghum plant with B deficiency. Note the shortened internodes.

45. Symptoms on sorghum leaves caused by excess nitrate-N added as calcium nitrate.

46. Symptoms on the second leaf of sorghum plants given excess ammonium nitrate.

47. Red speckling from P toxicity on first three leaves on sorghum plant grown in nutrient solution with 10.0 μM P (potassium dihydrogen phosphate) per plant.

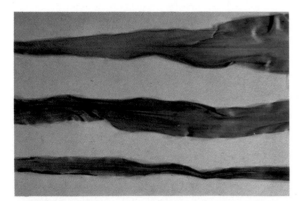

48. Leaves of sorghum plants grown in nutrient solution with excess KCl.

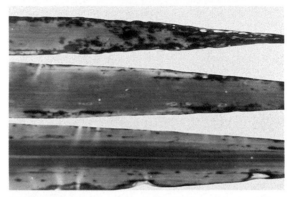

49. Leaves of sorghum plants grown in nutrient solution with excess CaCl.

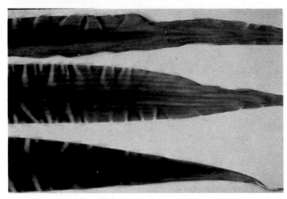

50. Leaves of sorghum plants grown in nutrient solution with excess MgCl.

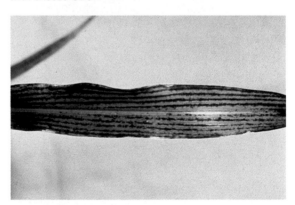

51. Advanced stage of toxicity on leaf of a sorghum plant grown in nutrient solution with excess Mn.

52. Leaves of sorghum plants grown in nutrient solution with excess Fe added as FeHEDTA.

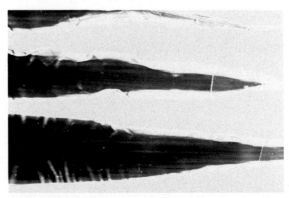

53. Leaves of sorghum plants grown in nutrient solution with excess B. Note sharp boundary between affected and unaffected tissue.

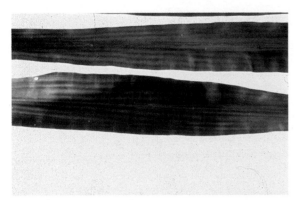

54. Leaves of sorghum plants grown in nutrient solution with excess Mo. Note similarity to P deficiency symptoms.

55. Leaves of sorghum plants grown in nutrient solution with Se. Note similarity to P deficiency symptoms.

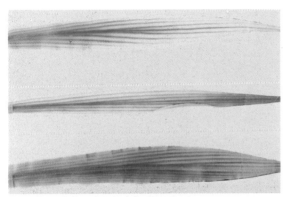

56. Symptoms typical of Fe deficiency on leaves of sorghum plants grown in nutrient solution with excess Al. Leaves with the most severe symptoms were emerging from the whorl, and other leaves were adjacent descending leaves on the plant.

57. Leaves of sorghum plants grown with different levels of Si. The yellow leaf is from a plant grown with high Si (7,200 μM); for the other leaves, decreasing levels of Si were added (bottom to top).

58. Leaves of sorghum plants grown in nutrient solution with high strontium.

59. Leaves of sorghum plants grown in nutrient solution with high Co. Note typical Fe deficiency streaking.

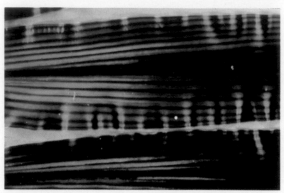

60. Leaves of sorghum plants grown in nutrient solution with excess Ni. Note typical Fe deficiency streaking.

61. Leaves of sorghum plants grown in nutrient solution with excess chromium. Note dull and diffuse orange color.

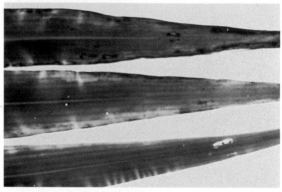

62. Leaves of sorghum plants grown in nutrient solution with excess lead. Note distinct orange color between green unaffected and affected tissue.

63. Leaves of sorghum plants grown in nutrient solution with excess cadmium. Note brilliant red of leaves with severe symptoms.

64. Wheat in nutrient-deficient soil showing localized growth responses to livestock excrement.

65. Winter wheat with N deficiency from leaching and denitrification (conversion of NO_3^- to gaseous N in the absence of oxygen) in wet soil.

66. Chlorosis of older leaves of wheat plants grown under conditions of deficient N.

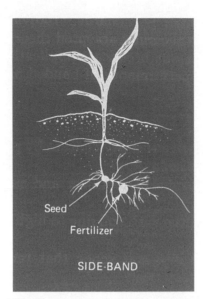

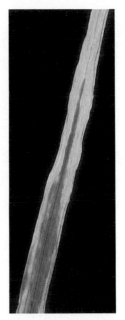

67. Phosphorus-deficient (right) and normal (left) wheat leaves.

68. Band application of nutrients such as P in the root zone to increase wheat plant uptake and utilization.

69. Potassium deficiency in an aging wheat leaf.

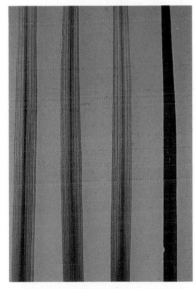

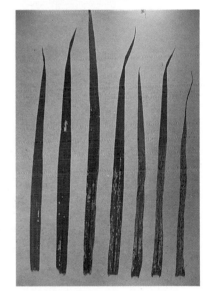

70. Young wheat leaves exhibiting interveinal Fe chlorosis; normal leaf on right.

71. Symptoms of Zn deficiency on leaves of oats.

72. Oat leaves spotted and discolored by Mn deficiency.

73. Leaf tips of wheat discolored and distorted from Cu deficiency.

74. Healthy wheat (left) and wheat head and leaves distorted by B deficiency (right).

75. Autumn-sown wheat developing in soil at pH 4.0 (left) and 6.0 (right).

76. Nitrogen-deficient sugarcane. Leaves become uniformly green to light yellow.

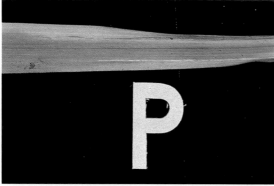

77. Dieback from tips and along margins of P-deficient older sugarcane leaves.

78. Potassium deficiency in sugarcane turns leaves brown ("fired" leaves).

79. Reddening of upper surface of leaf midrib in K-deficient sugarcane.

80. Red necrotic lesions from Mg deficiency in sugarcane.

81. Uniform chlorosis of S-deficient sugarcane.

82. Iron-deficient sugarcane with interveinal necrosis.

83. Chlorosis of entire sugarcane plant with Fe deficiency.

84. Zinc deficiency in sugarcane, with veinal chlorosis beginning at the base of young leaves.

85. Manganese deficiency in sugarcane, with interveinal chlorosis from the leaf tip to the middle.

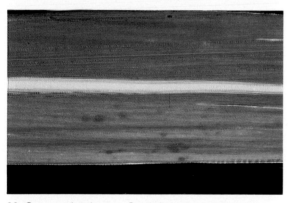

86. Green splotches on Cu-deficient sugarcane.

87. Translucent lesions between veins in B-deficient sugarcane.

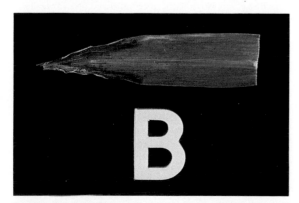

88. Burned leaf tip from B deficiency in sugarcane.

89. White freckles from Si deficiency in sugarcane.

90. Selection of young mature, fully developed leaf from typical sugar beet plant for petiole or blade analysis.

91. Sugar beet plants grown in N-deficient (left) and fertilized (right) soil.

92. Yellowing and wilting of older leaves of N-deficient sugar beet plant with small green center leaves.

93. Leaf sequence of N-deficient sugar beet plant with small green center blades (top); young mature green to light green to yellow blades (bottom). See also Plate 96.

94. Diphenylamine reagent test for nitrate. Positive (blue), from green young mature sugar beet leaf (left); negative (colorless), from green N-deficient center leaf (right). See Chapter 15 for reagent preparation.

95. Sulfur-deficient sugar beet plants with uniform yellowing (right); fertilized plants (left).

96. Uniform yellowing of young, young mature, and old leaves (left to right) of S-deficient (top) and normal green (bottom) sugar beet plants. See also Plate 93.

97. Positive blue diphenylamine test for nitrate in normal green (left) and S-deficient yellow (right) sugar beet leaves. The green and yellow blades contained nitrate-N at 2,500 and 9,000 ppm and sulfate-S at 1,800 and 150 ppm, respectively.

98. Uniform yellowing of Mo-deficient sugar beet leaf blade (right); normal blade (left).

99. Slight veining and pronounced necrotic spotting in sugar beet leaf from Mo deficiency.

100. Phosphorus deficiency, with stunted greening, of sugar beet plants (right); fertilized plants (left).

101. Phosphorus-deficient sugar beet plant with stunted greening (left); fertilized plant (right).

102. Old sugar beet leaf, P deficient, with brown veining (left); N-deficient leaf with uniform yellowing (right).

103. Potassium-deficient, low-Na sugar beet plant. Mature blades with crinkling and scorching are very low in K and Na; the corresponding petioles are high in "trapped" K and low in Na. Center green leaves, unscorched, are high in K.

104. Potassium-deficient, high-Na sugar beet plant. Blades and petioles of scorched mature leaves are low in K and high in Na. Green center leaves are high in K.

105. Potassium-deficient, mature sugar beet leaf blades without (left) and with (right) Na addition. Sodium increases blade size and greatly reduces crinkling and scorching.

106. Magnesium-deficient sugar beet plant with scorched mature leaves, often retaining a green arrowhead pattern at the leaf blade base.

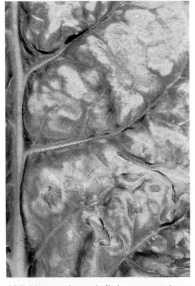

107. Severe chlorosis and necrosis of mature sugar beet leaf with Mg deficiency.

108. Magnesium-deficient sugar beet leaf showing green veins, yellow interveinal areas with green islets, and start of necrosis.

109. Calcium-deficient sugar beet plant with severe tipburn of center leaves and crinkled, deformed blades of older leaves.

110. Calcium-deficient sugar beet leaves with black-tipped, hooded, and deformed leaf blades.

111. Cross section of Ca-deficient sugar beet storage root. Dark rings are damaged cambium and conducting tissues.

112. Crinkled, cracked, sometimes syrupy, upper blade surface of B-deficient sugar beet leaf.

113. Boron-deficient sugar beet plant with dark green crinkled, deformed, and, at times, wilted center leaves.

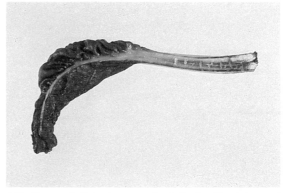

114. Boron-deficient sugar beet leaf with deformed blade and netted cracking of petiole.

115. Iron-deficient sugar beet plant with uniform yellowing of young leaves.

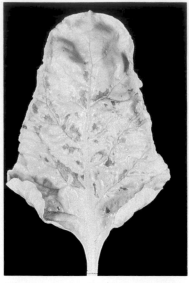

116. Severe Fe-deficient sugar beet leaf with yellowing, bleaching, and necrosis.

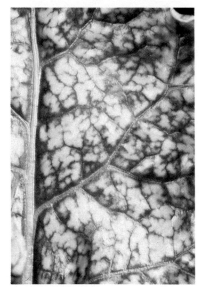

117. Pronounced green veining, a symptom of Fe recovery, in sugar beet leaf.

118. Zinc-deficient sugar beet plant with yellowing followed by pitting, interveinal necrosis, and pronounced green veining.

119. Initial pitting of mature sugar beet leaf blade with Zn deficiency.

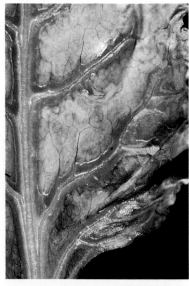

120. Interveinal necrosis and prominent green main veins of Zn-deficient mature sugar beet leaf blade.

121. Manganese-deficient sugar beet plant with yellowing leaves developing a metallic sheen and black spotting.

122. Dark, freckled, necrotic areas of Mn-deficient sugar beet leaf.

123. Black-pigmented borders along main veins of sugar beet leaf blade with severe Mn deficiency.

124. Chlorosis of young center sugar beet leaves with Cu deficiency.

125. Green veining and start of bleaching in Cu-deficient sugar beet leaf.

126. Bleaching caused by Cu deficiency in sugar beet leaf.

127. Chlorine deficiency in sugar beet, first appearing as a chlorosis of the young center leaves.

128. Chlorine-deficient sugar beet blade with raised veins and interveinal chlorosis (left); normal blade (right).

129. Nitrogen deficiency in soybean. Plants turn uniformly pale green, then yellow.

130. Potassium deficiency in soybean, producing mottling that becomes an irregular yellow border.

131. Potassium deficiency in soybean, with necrosis of chlorotic areas of leaflets.

132. Ridging of interveinal tissues in Ca-deficient soybean.

133. Excess Mg (left) and Mg deficiency (right) in soybean.

134. Interveinal yellowing of younger leaves of Fe-deficient soybean.

135. Zinc deficiency in soybean.

136. Manganese deficiency in soybean.

137. Newly opening leaves of soybean with B toxicity.

138. Manganese toxicity in soybean.

139. Nitrogen deficiency in peanut.

140. Nitrogen deficiency in peanut.

141. Potassium deficiency in peanut.

142. Calcium deficiency in peanut, commonly resulting in abnormal fruiting.

143. Calcium-deficient peanuts, with darkened plumules.

144. Magnesium deficiency in peanut.

145. Green veins and yellow interveinal tissues of Fe-deficient peanut leaves.

146. Stem lesions on peanut caused by Zn toxicity.

147. Chlorotic interveinal areas of severe Mn deficiency in peanut.

149. Molybdenum deficiency in peanut.

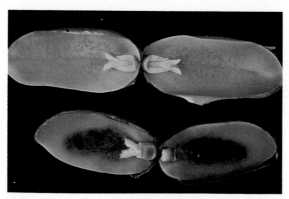

148. Hollow heart of peanuts (bottom) from B deficiency.

150. Early-season N deficiency symptoms in cotton on a three-row strip (right) where no N fertilizer was applied.

152. Interveinal chlorosis from early-season K deficiency in cotton, which first appears on older leaves.

151. Leaf reddening symptoms of late-season P deficiency on Acala cotton.

153. Foliar symptoms of late-season K deficiency on cotton leaves in upper third of canopy.

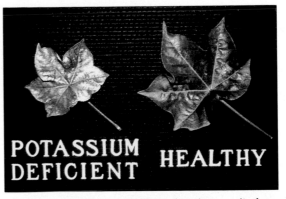

154. Close-up detail of leaf bronzing that results from late-season K deficiency in cotton.

155. Variation in leaf discoloration due to late-season K deficiency in cotton; normal leaf at center. Leaf at lower right shows marginal necrosis without defined borders.

156. Interveinal necrotic lesions with well-defined borders from Verticillium wilt infection in cotton (left); leaf bronzing symptoms from late-season K deficiency (right).

157. Interveinal purple-red discoloration from Mg deficiency in cotton.

158. Sulfur deficiency in cotton, appearing on upper leaves as pale to yellowish green discoloration.

159. Distinct interveinal chlorosis caused by a deficiency of Zn in cotton; leaf veins remain green.

160. Crinkle leaf symptoms caused by Mn toxicity in cotton.

161. Cucumber exhibiting symptoms of N deficiency.

162. Potassium deficiency in cucumber, with mottled chlorosis on leaf tips of older leaves that become necrotic.

163. Magnesium deficiency in cucumber. Older leaves are chlorotic beginning at leaf tips and between veins.

164. Copper deficiency in watermelon. Note crinkled leaves and compact appearance on a severely Cu-deficient 3-month-old plant.

165. Symptoms of N deficiency in wet areas of an onion field after a heavy rainfall.

166. Onion plants P broadcast (left) and P banded 2″ below and to the side (right) (N and K on both).

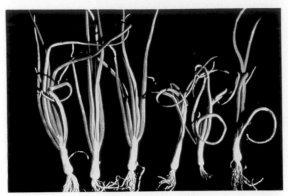

167. Zinc-deficient (right) and normal (left) onion plants.

168. Magnesium-deficient (right) and normal (left) onion plants.

169. Onions grown in organic soil amended with copper sulfate (left). White leaf tips and spiral or right-angle twisting typical of Cu deficiency in unamended soil (right).

170. Copper-deficient onion bulbs with thin pale yellow outer scales (right) and normal bulbs (left).

171. Dieback and wilting of leaves of Mo-deficient onions.

172. Center row of onions, in organic soil, stunted by Mo deficiency. Outside rows sprayed with sodium molybdate.

173. Light green tomato leaves caused by N deficiency.

174. Purple interveinal tissue on underside of leaf of tomato seedling caused by P deficiency.

175. Dark green leaves of tomato seedling with young puckered upper leaves caused by K deficiency.

176. Necrotic spotting of interveinal tissue of older tomato leaves from K deficiency.

177. Limited fruit flesh development in K-deficient tomatoes (right).

178. Upward cupping of leaves, marginal necrosis, and terminal tissue in Ca-deficient tomato plant.

179. Dead tomato root tip, with roots branching behind, from Ca deficiency.

180. Blossom-end rot physiologically caused by Ca deficiency in tomato.

181. Yellowing and white chlorotic and necrotic interveinal tissue of older tomato leaves caused by Mg deficiency.

182. Tip tissue chlorosis in tomato on high-pH sand hydroponic system. Free $CaCO_3$ induces Fe deficiency.

183. Stunting of Zn-deficient tomato plant with white and necrotic interveinal spotting of older leaves.

184. Interveinal chlorosis, with green veins, of expanded young tomato leaves due to Mn deficiency.

185. Boron deficiency in tomato induced by excess lime on tropical soil (pH > 7). Note severely stunted seedling.

186. Stunting and whitening of developing leaves caused by Al toxicity in tomato plant.

187. Necrotic margins of older tomato leaves caused by B toxicity at hydathodes.

188. Veinal necrosis and leaf yellowing due to Mn toxicity in tomato.

189. Marked shortening of internodes of normal-appearing bean plant, typical of Zn toxicity in tomato.

190. White, green-veined tomato seedling foliage (interveinal chlorosis), an indication of NH_3 toxicity.

191. Interveinal chlorosis of older and general chlorosis of younger bean leaves caused by Fe deficiency.

192. Iron deficiency symptoms in bean induced by temporary flooding of field.

193. Interveinal chlorosis and necrosis of bean leaves caused by Zn deficiency.

194. Early symptoms of interveinal chlorosis on young bean leaflet caused by Mn deficiency.

195. Advanced symptoms of interveinal chlorosis on young bean leaves caused by Mn deficiency.

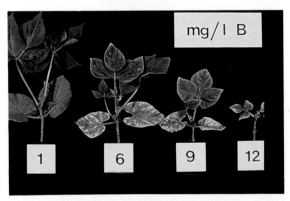

196. Yellowing, necrosis, and stunting of bean plants caused by excess B.

197. Sturdy, vigorous, nondeficient potato plant.

198. Young mature, nondeficient potato leaf.

199. Healthy terminal potato leaf blade.

200. Uniformly light green leaves of mildly N-deficient potato plant.

201. Severe N deficiency in potato, with upward cupping of leaf blades. Results in negative quick test for nitrate with diphenylamine reagent on petiole (see Plate 232).

202. Uniform yellowing of N-deficient terminal potato blade.

203. Light green younger leaves of S-deficient potato plant.

204. Decreasing S deficiency (left to right) of youngest mature potato leaves. Note uniform yellowing and mild upward curling of deficient leaf blades.

205. Molybdenum deficiency in potato, with symptoms similar to those of N and S deficiency.

206. Phosphorus deficiency in potato. Note dark green color and stunted growth.

207. Mild P deficiency in potato. Note dark green color and mild leaf roll.

208. Phosphorus deficiency in potato. Note mild to severe leaf roll and gray-green lower leaf surface.

209. Potassium-deficient potato plant, with glossy sheen, pronounced crinkling, and marginal necrosis on young mature leaves. Crown leaves remain green.

210. Severe K deficiency in potato leaf, with pronounced crinkling, marginal leaf scorch, and slightly black pigmentation.

211. Marginal leaf scorch of severely K-deficient potato leaf.

212. Magnesium deficiency in potato. Begins on young mature leaves as slight chlorosis with green veining and brown spotting. Crown leaves remain green.

213. Magnesium deficiency in potato, with green veining and interveinal browning.

214. Severe chlorosis and leaf scorch of Mg-deficient potato leaf.

215. Calcium deficiency in potato. Begins on youngest leaves; with severe deficiency, young mature leaves droop with chlorosis and brown spotting, and blades cup upward.

216. Calcium-deficient young mature potato leaf. Blades cup upward with chlorosis and brown spotting.

217. Calcium deficiency in potato. Growing points of terminal and lateral shoots wilt and die. Root tips swell and darken (not shown).

218. Boron deficiency in potato. Note bushy, droopy leaves and crinkled, upward-cupping blades bordered by light brown dry tissue.

219. Mature leaf of B-deficient potato plant. Note browning of blade edges and veins prior to drying.

220. Boron deficiency in potato. Blades of center leaves become crinkled, cup upward, and are bordered by light brown tissue; the growing point dies. Root tips swell and darken (not shown).

221. Iron deficiency in potato. Young new leaves near growing point of plant become yellow to white or chlorotic, usually without necrosis.

222. Iron-deficient potato potato plant, with yellow young crown leaves. Blade tips remain green longest.

223. Iron deficiency in potato. Note yellowing and green veining, often a recovery sign.

224. Zinc deficiency (little leaf or fern leaf) in potato. Young leaves develop chlorosis and form narrow, upward-cupped leaf blades with tipburn.

225. Severe, moderate, and mild symptoms of Zn deficiency in potato.

226. Potato leaf blade with severe, moderate, and mild symptoms of Zn deficiency.

227. Manganese deficiency in potato, beginning with yellowed and slightly cupped younger leaves with brown to black spots between veins.

228. Very severe Mn deficiency in potato. Black spotting (left); yellowing, cupping, and brown to black spotting on terminal blade and very severe symptoms on lateral blades (center); and mild symptoms on leaf (right).

229. Cupping and brown and black spotting on severely Mn-deficient potato leaf blade.

230. Early stage of Cu deficiency in potato, not readily detectable on young blades.

231. Severe Cu deficiency in potato. Note upward cupping and inward rolling of young blades.

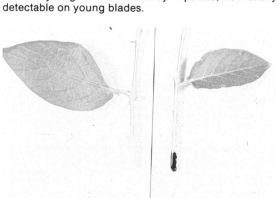

232. Diphenylamine quick test for nitrate. Negative test (colorless, left) indicates N-deficient potato plant. Instant deep blue color of positive test (right) indicates yellowing is due to S or Mo deficiency or to other causes.

233. Nitrogen deficiency in Bartlett pear.

234. Phosphorus deficiency in apple.

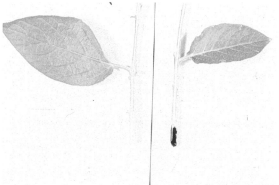

235. Phosphorus deficiency in D'Anjou pear.

236. Potassium-sufficient apple fruit (left); fruit with poor coloration associated with inadequate K (right).

237. Calcium deficiency (bitter pit) in Northern Spy apple fruit.

238. Magnesium deficiency in apple.

239. Magnesium deficiency in pear.

240. Sulfur-deficient (right) and normal (left) apple shoot.

241. Iron deficiency in Ana apple.

242. Zinc-deficient Red Delicious apple shoots.

243. Zinc-deficient Granny Smith apple shoots.

244. Manganese deficiency in apple.

245. Wither tip (Cu deficiency) on young apple.

246. Blossom blast (B deficiency) in pear.

247. Boron-deficient pear fruit.

248. Bark necrosis (B deficiency) on Red Delicious apple.

249. "Measles" (Mn toxicity) on Red Delicious apple.

250. Nitrogen-deficient citrus leaves showing yellowing accompanying N withdrawal from leaves subtending new growth.

251. Effect of N recycling over 18-month growth period on grapefruit twigs from trees maintained on low (left) and adequate (right) N. **A,** Green terminal leaves. **B,** Yellowing. **C,** Defoliation.

252. Large, puffy, P-deficient orange fruit with thick peel (right); healthy fruit (left).

253. Potassium-deficient grapefruit leaves with green mottling pattern in distal portion of leaves.

254. Valencia oranges from trees maintained with levels of leaf K ranging from 1.0 (left) to 2.0% (right).

255. Mottling along margins of Ca-deficient orange leaves.

256. Bronzing pattern on newly matured Mg-deficient orange leaves.

257. Iron-deficient young orange leaves, with green veins and midrib.

258. Strong chlorosis from Zn deficiency in small attenuated orange leaves.

259. Mixed chlorotic pattern in orange leaves deficient in both Zn and Mn.

260. Full-size orange leaves with light and dark green mottle pattern from Mn deficiency.

261. Elongated orange leaf and twig dieback from Cu deficiency.

262. Young orange twigs with gum pockets from Cu deficiency.

263. Dull, lusterless leaf color on orange shoot from B deficiency (right).

264. Young B-deficient orange fruit with gum deposits in peel and premature development of rind color.

265. Yellow spots from Mo deficiency in mature orange leaves.

266. Salinity damage to orange leaves with sulfate and chloride toxicity.

267. Orange leaves with tip and marginal necrotic spots from Na toxicity.

268. Chlorine toxicity in grapefruit, with characteristic bronzing of leaves in its early stages.

269. Boron toxicity in grapefruit leaves, with irregular bronzing chlorosis mostly in distal part of leaf.

270. Stone fruit leaves with red spots characteristic of N deficiency.

271. Severe N deficiency in stone fruit, with red coloration on leaves and stem.

272. Potassium-deficient peach tree in late summer.

273. Inverted V pattern typical of Mg-deficient stone fruit leaves.

274. Sulfur-deficient peach shoot (right), with uniformly yellow leaves at tip.

275. Iron-deficient stone fruit plant, with green veins and interveinal chlorosis.

276. Little-leaf rosetting in Zn-deficient peach (left).

277. Severe Zn deficiency in stone fruit. Basal leaves abscise, leaving tufts of small, pointed leaves at shoot tips.

278. Characteristic herringbone pattern of Mn-deficient stone fruit leaves.

279. Long, narrow leaves and interveinal chlorosis of Cu-deficient stone fruit.

280. Boron-deficient stone fruit tree, with twig dieback and small leaves.

281. Chloride toxicity in stone fruit leaf (left), showing marginal leaf burn; normal leaf (right).

282. Nitrogen deficiency in grape, with pink to reddish shoot and petioles. All blades are smaller, thinner, and yellower than normal.

283. Advanced-stage early spring symptoms of K deficiency in grape, with distorted, ruffled blades with sporadically distributed marginal and interveinal necrosis.

284. Summer symptoms of K deficiency in grape, with shining spots between main veins caused by partial desiccation of epidermis ("nail enamel" spots or "oil stains"). After several days, initially yellow spots turn yellowish brown and, finally, lilac brown.

285. Autumn symptoms of K deficiency in grape, with lilac to dark brown epidermis (black leaf), often spreading over entire leaf surface. Margins of blades die off, dry up, and roll upward.

286. Strong marginal burn from Ca deficiency in grape leaves, with necrosis advancing from edge toward center of leaf blade and sporadically distributed dark brown pimples on shoot bark.

287. Summer and autumn symptoms of Mg deficiency in grape. Note wedge-shaped yellowed (or reddish, on red varieties) areas between the main veins and marginal necrosis on aged leaves. See also Figure 19.2.

288. Acidity damage associated with Ca and Mg deficiency in grape. Several weeks after bloom, margins of older leaves become yellowish to light brown, followed by advancing desiccation of borders.

289. Early-stage stem necrosis from Ca and Mg deficiency in grape, with shallow and concave desiccations of cortex on main axis of rachis shortly after beginning of ripening.

290. Lime-induced chlorosis in Fe-deficient grape, with yellowing of leaf blade tissue between veins, which remain green.

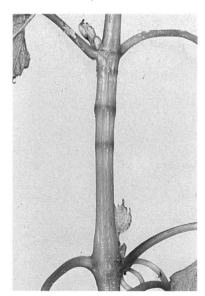

291. Iron deficiency in grape. Shoots have lost their leaves; lateral shoots with small, narrow, folded, completely bleached leaves arise from axillary buds. Tendrils are large in relation to leaves. Lateral shoot bases are reddish at point of attachment.

292. Zinc deficiency in grape. Note small leaf with sharp teeth, open petiolar sinus, lightly protruding veins, and asymmetric blade with unequal areas on left and right sides of midrib.

293. Manganese deficiency in grape, with mosaiclike disposed yellow area bounded by fine veins in interveinal areas.

294. Boron deficiency in grape. Note internode with two pronounced and one weak dark green nodelike intumescences (swellings). See also Figure 19.4.

295. Advanced-stage B deficiency in grape leaves, with interveinal chlorosis and necrosis.

296. Coralloid deformed roots of B-deficient grape, with longitudinal lacerations on swollen axis and club-shaped tips of lateral rootlets, some with necrotic apexes.

297. Boron excess in grape leaf blades, with toothless, downward-rolled margins and shallow, dark brown necrotic spots between veins. Lobes of leaf appear rounded.

298. Strong burn of leaf blade margins from Cl excess in grape, induced by sprays with aqueous solutions of Mg or NaCl to prevent stem necrosis.

299. Marginal and interveinal necrosis in grape caused by hydrogen fluoride emitted from an aluminum plant.

300. Nitrogen deficiency and excess N on same turf-grass area due to improper spreader application.

301. Phosphorus-deficient plot (left) six weeks after seeding Kentucky bluegrass.

302. Early signs of P deficiency on mature creeping bentgrass. Turf darkens and has appearance of mild drought stress.

303. Advanced P deficiency on mature creeping bentgrass. Stand is reddish purple due to deficiency symptoms on older leaves.

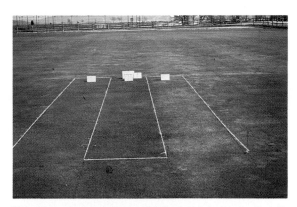

304. Early spring chlorosis (right strip) due to K deficiency on creeping bentgrass.

305. Chlorosis on bluegrass due to Fe deficiency, showing typical irregular pattern.

306. Interveinal chlorosis caused by Fe deficiency in turfgrass.

307. Blackening of turfgrass from Fe toxicity caused by foliar Fe application.

Chapter 7

Albert Ulrich
Department of Soil Science
University of California, Berkeley

J. T. Moraghan
Department of Soil Science
North Dakota State University, Fargo

E. D. Whitney
USDA Agricultural Research Service
Salinas, California

Sugar Beet

The commercial production of beet sugar has been an outstanding achievement scientifically and economically as an alternate source of sugar when other supplies are insecure. What needs to be told now more precisely is how to meet the fertilizer requirements of sugar beet crops being grown under the widely differing soil types, cultural practices, and climatic conditions of the temperate zone. The first of five steps to be taken to become more efficient in fertilizer usage, particularly with nitrogen, is to ask the plant about its fertilizer needs by comparing the observed deficiency symptoms with those illustrated in the plates accompanying this chapter. The next four steps will anticipate and prevent nutrient deficiencies in subsequent crops.

Agronomically, the sugar beet is one of the four cultural types of *Beta vulgaris* L.: sugar beet, red beet or garden beet, Swiss chard, and fodder beet. The fleshy root is the storage organ of the sugar beet and is the source of commercial sugar and also of wet and dry beet pulp, which serves as animal feed. At a young stage of growth, the tops of all cultural types, especially Swiss chard, may serve as a leafy vegetable, and in later stages of growth as livestock forage. The fleshy root and tops of the red beet are an excellent vegetable.

Botanically, the sugar beet is classified as a biennial plant that produces leaves the first year until the plant becomes dormant during the winter. Growth restarts in the spring to form new leaves and a shoot, which produces seed by early summer. This stage is important to survival of the plant, but to the sugar beet grower it is the first year's growth that is important for sugar production. During the first year of vegetative growth, the sugar beet has no internal mechanism for "ripening" to accumulate sugar in the storage root like the ripening of the tomato fruit. However, a false ripening process of the sugar beet can occur, primarily through external forces such as during periods of low night temperature and/or of N deficiency.

Sugar beet plants thrive best in the temperate zone on a deep well-drained loam soil rich in mineral nutrients, including N, which is the key element for vigorous top and root growth. As the taproot enlarges, sucrose is stored within the parenchyma cells of the storage root at an equilibrium concentration characteristic of the cultivar selected for sugar production. Under high-N conditions, most commercial cultivars have a sucrose concentration from 6 to 10%, which may reach 18–20% for a sugar-type. These equilibrium values increase moderately during the cool, sunny weather of autumn, and when combined with N depletion for 4–6 weeks prior to harvest, the sugar concentrations often increase dramatically. Under these conditions, beets of medium size with N deficiency in a uniform stand will ripen rapidly, often to 18% sugar, to produce an easily processed quality crop.

Sugar accumulation is an inverse relationship between sucrose concentration in the storage root and N uptake. Thus, there is a need for a judicious reduction of N fertilizer to shift the sugar beet from the vegetative stage of growth to the sugar storage phase just prior to harvest. Formulae have been developed to estimate N needs on the basis of anticipated yield, N required per ton of beets, the mineralizable soil N, preplant level of soil nitrate-N, and seasonal petiole nitrate-N patterns, with these being referenced to critical and safe N levels (Fig. 7.1). If irrigation water is high in nitrate-N, this source of plant N must be taken into account.

In producing a quality crop of beets for sugar production, it is necessary for the grower to plant a superior cultivar of beet, developed to grow rapidly and be resistant to disease from planting to harvest. Sugar beet growers need to plant early to a uniform stand, meet the water and fertilizer requirements of the crop, prevent weed competition, and control pests.

Deficiency Symptoms

Uniform Yellowing

Nitrogen (N)

Two distinctive changes in leaves characterize N deficiency (Plates 90–94).

Overall yellowing of leaves occurs when a plant first becomes N deficient. The plant may previously have had high levels of N and may have grown vigorously to form large leaves of a good green color (Plate 90). With nitrate depletion, however, these leaves become light green, turning to yellow. Yellowing continues as the plant ages, accompanied by wilting and an accelerated death rate of the older leaves (Plates 91–93).

While these changes are taking place in the large leaves, other N deficiency symptoms occur that are often overlooked. Newly formed leaves in the center of the plant are much smaller and narrower than older leaves and turn an intense green (Plates 91 and 92). Leaves often lie nearly parallel to the soil surface, with the petioles curved slightly upward. The size of the center leaves depends on the rate at which nitrate is formed in the soil and absorbed by the plants. If this rate is low, the center leaves are small and lie close to the soil surface. If the rate of nitrate supply is fairly high but still insufficient in the petioles, the new leaves are larger, perhaps almost large enough to cover the soil surface between rows and make the field appear uniformly green. Only a close inspection of plants and the diphenylamine quick test (see the Appendix to Chapter 15 and Plate 232) for plant nitrate can reveal immediately if the yellowing is due to some factor other than N deficiency (Plate 94). A side-dressing of 80 lb/acre of N (90 kg/ha) from ammonium nitrate or some other readily available N source before harvest will correct the N deficiency. Nitrogen deficiency is indicated under field conditions when petioles from recently matured leaves as shown in Plates 91–94 contain less than 500 ppm of nitrate-N.

Sulfur (S)

Early symptoms of S deficiency are similar to those of N deficiency. Leaves of entire plants change gradually from green to light green (Plate 95), then to light yellow with a faint tinge of green remaining (Plate 96). Veining is not a prominent feature of either N or S deficiency. In time, however, the symptoms of S deficiency gradually become different from those of N deficiency. The new center leaves of S-deficient plants become light green to almost yellow (Plate 96) rather than the deep green

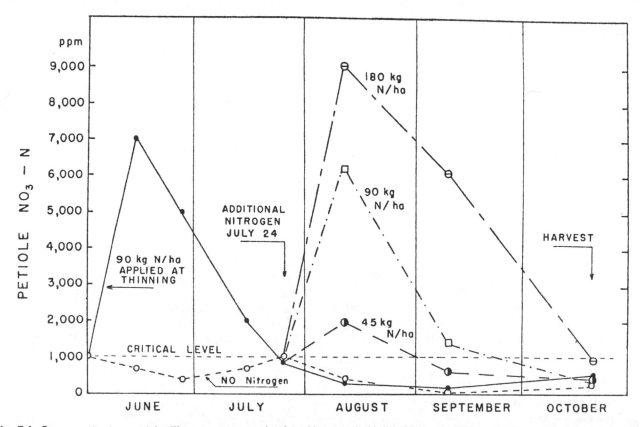

Fig. 7.1. Crop monitoring and fertilizer program evaluation. Untreated; 80 lb of N/acre (90 kg of N/ha) side-dressed as NH₄NO₃ at thinning on May 31; plus additional N at 40, 80, and 160 lb/acre (45, 90, and 180 kg/ha) applied on July 24 when petiole samples indicated the plants would soon be deficient in N. Sugar yields on October 20 were 6,680, 7,280, 7,820, 8,400, and 8,480 lb/acre (7,490, 8,160, 8,770, 9,420, and 9,510 kg/ha). The highest economical return was obtained from the second application of 80 lb of N/acre (90 kg/ha) applied in mid-July. From research near Davis, California. Reprinted, by permission, from Ulrich et al (11).

of N deficiency (Plate 93). Unlike N deficiency, the leaves of S-deficient plants remain erect as their center leaves change from green to yellow; the petioles and blades become brittle and break readily if compressed by the hand. In cases of extreme deficiency, brown blotches form in irregular patterns on leaf blades, without regard to venation or location, and may later coalesce to form larger blotches. Brown blotches may also appear on petioles, but they generally do not become necrotic.

Symptoms of S deficiency are likely to occur under field conditions where sulfur dioxide is low as a pollutant, irrigation water is very low in sulfate-S, and ammonium sulfate and S-containing manures have been replaced by fertilizers without S as sources of N for crops including sugar beets.

Sulfur-deficient petioles are usually high in nitrate. Thus, the diphenylamine test can rule out the possibility that the yellowing of leaves is caused by N deficiency. If the reagent turns blue to deep blue-black on the cut surface of the petiole of a yellow leaf (Plate 97), the yellowing is quite likely due to S deficiency. A side-dressing of gypsum at the rate of 2,000 lb/acre (2,250 kg/ha) will correct the S deficiency. Sulfur deficiency is indicated when leaf blade tissue contains less than 250 ppm (dry weight basis).

Molybdenum (Mo)

Symptoms of Mo deficiency first appear as a general yellowing (Plate 98), similar to that caused by S deficiency or, to some extent, by N deficiency. The center leaves are light green to yellow, as with S deficiency, in marked contrast to the deep green of N deficiency. As the symptoms of Mo deficiency increase in severity, pitting develops along leaf veins (Plate 99). This symptom differs distinctly from the black sheen and spotting of Mn deficiency and the coalescing of spots in Zn and S deficiencies. The diphenylamine reagent shows high nitrate content in the petioles and leaf blades of Mo-deficient plants, just as it does in S-deficient plants. Mo deficiency has not been observed in commercial fields. Blades with deficiency symptoms contain less than 0.20 ppm of Mo (dry weight basis).

Stunted Greening

Phosphorus (P)

Phosphorus deficiency is by far the most difficult deficiency symptom to recognize. An overall stunting of the plant and a gradual deepening of the green color of foliage are the only visible symptoms (Plates 100 and 101). Except for their size, such plants appear normal and look as if they had been planted several weeks later than high-P plants of the same age. As the deficiency becomes more severe, the deep green color often assumes a metallic luster ranging from dull grayish green to almost bluish green. The purpling that is often associated with P deficiency in other crops seldom appears on sugar

beet and is not a reliable criterion for diagnosis. In the early seedling deficiency stage, cotyledons and primary leaves are deep green; if the deficiency is severe, pitting occurs, followed by drying of the cotyledons. In severely deficient plants, brown netted veining forms in the tissues of older leaves when they dry, in contrast to the uniform yellowing that occurs on normal and N-deficient plants (Plate 102).

Phosphorus deficiency occurs frequently in seedlings when soil temperatures are 50°F (10°C) or lower. This deficiency can be prevented by applying P at planting to one side and just below the seed level. Also, P deficiency may occur on peat, calcareous, acid, and kaolinitic soils. Since the symptoms are not readily recognizable, plant samples should be taken shortly after emergence, just after thinning, and after that at 4-week intervals to harvest. Phosphorus deficiency is indicated when petioles or blades contain less than 750 ppm of H_2PO_4-P soluble in 2% acetic acid.

Leaf Scorch

Potassium (K)

The first sign of K deficiency appears as a tanning and leathering of the edges of recently matured leaves and thereafter differs according to the sodium (Na) status of the plant. When the soil or nutrient solution is very low in Na, a severe interveinal leaf scorch and crinkling proceeds to the midrib. During this time the tissues bordering the midrib and veins remain green (Plates 103 and 104). In contrast, under relatively high Na conditions, tanning and leaf scorch lead to a smooth leaf surface (Plates 104 and 105). In both instances, the K concentration of the blades ranges narrowly from 0.2 to 0.6%, but for the low-Na plants with petioles below 1.5% Na the petioles have high K values up to 2% K (dry weight basis) (Table 7.1). Absorption of Na at this time causes the "trapped" petiole K to move to the blade, where it increases blade growth under low-K conditions (Plate 104). When soil available K is low and soil nitrate-N abundant, Na accompanies NO_3-N in the absorption process to increase storage root Na, thereby lowering beet quality. The increased top growth increases sugar production. In this way, Na substitution for K is a mixed blessing. Sodium also increases sugar beet growth even under high-K conditions when grown hydroponically. In this scenario the young center leaves and old outer leaves without symptoms are high in K. Leaf blades with pronounced deficiency symptoms often emit a faint putrescine odor. Potassium deficiency may occur in plants on sandy loam soils with low exchange capacity, on clay loam soils with a high unavailable K fixing power, on serpentine soils high in exchangeable Mg, or on peat, muck and other soils releasing K too slowly during the grand growth period of the crop. Potassium deficiency is indicated whenever blades contain less than 1.0% K.

Magnesium (Mg)

Magnesium deficiency (Plate 106) often resembles K deficiency in that the margins of recently matured leaves develop an interveinal chlorosis (Plate 107) towards the distal end. This becomes a more intense lemon-yellow color than that found with K deficiency. These areas gradually enlarge. Portions of leaves adjacent to the midrib and principal veins remain green for extended periods. When the deficiency is severe, chlorotic zones gradually become necrotic (Plate 108). These necrotic areas have a lighter brown color than those resulting from K deficiency. A characteristic green triangular area is sometimes observed near the leaf base (Plate 106). Magnesium deficiency may be confused with "virus yellows." This may occur when the chlorotic zones in virus yellows become a bright yellow with orange tints. Also, leaf thickening is more prominent in plants with virus yellows. Magnesium deficiency occurs frequently on acid soils that have been extensively leached or on soils fertilized frequently with K. Magnesium deficiency is indicated when petioles from leaves with symptoms contain less than 0.03% or blades less than 0.05%.

Growing-Point Damage

Calcium (Ca)

Crinkling and downward cupping, or hooding, of young leaf blades are the first symptoms of Ca deficiency (Plates 109 and 110). As these leaves develop, the blades crinkle and fail to develop to full size, because Ca is

Table 7.1. Plant analysis guide to determine nutrient deficiency and nondeficiency in sugar beet[a,b]

Nutrient constituent determined	Plant part tested[c]	Critical concentration level[d]	Range showing symptoms	
			Deficiency	No deficiency[e]
Boron, ppm	Blade	27	12–40	35–200
Calcium, %	Petiole	0.1	0.04–0.10	0.2–2.50
	Blade	0.5	0.1–0.4	0.4–1.5
Chlorine, %	Petiole	0.4	0.01–0.04	0.0–8.5
Copper, ppm	Blade	...	<2	>2
Iron, ppm	Blade	55	20–55	60–140
Magnesium, %	Petiole	...	0.010–0.030	0.10–0.70
	Blade	...	0.025–0.050	0.10–2.50
Manganese, ppm	Blade	10	4–20	25–360
Molybdenum, ppm	Blade	...	0.01–0.15	0.20–20.0
Nitrogen (N)				
NO_3^--N, ppm	Petiole	1,000	70–200	350–35,000
	Blade	300	0–400	500–11,700
Total N, %	Petiole	1.6	1.1–1.5	1.6–4.25
	Blade	2.5	1.9–2.9	3.0–6.00
Phosphorus (P)				
$H_2PO_4^-$-P, ppm	Petiole	750	150–400	750–4,000
	Blade	...	250–700	1,000–8,000
Total P, %	Petiole	1,200	600–1,200	1,400–11,000
	Blade	1,800	1,000–1,750	2,500–12,500
Potassium, %	Petiole[f]	1.0	0.2–0.6	1.0–11.0
	Blade	1.0	0.3–0.6	1.0–6.0
Sodium, %	Petiole[f]	...	...	0.02–9.0
	Blade	...	...	0.02–3.7
Sulfur (S)				
SO_4^--S, ppm	Blade	250	50–200	500–14,000
Total S, ppm	Blade	750	400–750	800–3,000
Zinc, ppm	Blade	9	2–13	10–80

[a] All values are based on dry weight.
[b] Leaf material for analysis must be collected shortly after symptoms appear. If this precaution is not taken, deficient plants may accumulate nutrients within the leaf without restoring "dead" tissues.
[c] Unless otherwise designated, blades and petioles are from a recently matured leaf (11).
[d] That concentration of a nutrient at which growth of a plant is retarded by 10%.
[e] The upper value reported is the highest observed to date for "normal" plants. Abnormally high values are often associated with other nutrient deficiencies; e.g., blades low in Fe may contain up to 4% Ca, 900 ppm Mn, and the like.
[f] Because Na affects K content in petioles, blades must be used for K analysis when petioles contain less than 1.5% Na.

necessary for blade expansion. In cases of severe deficiency, the blades become merely stubs of blackened tissue at the ends of the petioles (Plates 109 and 110). These symptoms, often called "tipburn," occur in fields where plant growth is extremely vigorous under high N and the cool weather conditions of spring. The symptoms even appear on plants growing on calcareous soils and in nutrient solutions containing up to 50 ppm of soluble Ca. However, without Ca in the root environment, roots fail to grow. Fortunately, tipburn usually disappears under field conditions as soil N decreases. Furthermore, tipburn has no material effect on yields or on sugar concentration, although at times the growing point is damaged and storage root rot may occur (Plate 111). Corrections usually include liming of acid soils and adding gypsum to neutral soils. Calcium deficiency of sugar beets is indicated primarily by symptoms or when Ca is less than 0.10% in petioles and 0.40% in blades.

Boron (B)

The first symptom of B deficiency is a white, netted chapping of the upper surface of leaf blades (Plate 112) or wilting of tops, even in plants growing in moist soil or in an aerated culture solution (Plate 113). Young leaves wilt sooner than older leaves (in a true water shortage, the young leaves wilt last). In time, the young leaves collapse and fail to develop (Plate 113).

Other symptoms are pronounced crinkling of leaf blades and darkening and cracking of petioles (Plate 114). Also, the growing point dies, and the crown darkens and becomes subject to rot or decay (Plate 113). Internal and external crown darkening symptoms are easily identified in the field.

Appearance of B deficiency symptoms in late June alerts the grower to the possibility of a B problem in future years. Unless local experience indicates otherwise, try small-scale field trials with sodium borate (ordinary borax) applied at the rate of 10–15 lb/acre (11–17 kg/ha) preplant. Later applications as a side-dressing or as foliar sprays after symptoms have appeared have not been effective in preventing yield reductions from B deficiency. Extreme care must be taken not to apply B fertilizer indiscriminately because of the danger of inducing B toxicity in other crops in the rotation. Boron deficiency is indicated as illustrated in Plates 112 and 113 and confirmed when blades with symptoms contain less than 30 ppm of B.

Yellowing with Interveinal Chlorosis

Iron (Fe)

Iron deficiency chlorosis in sugar beet can be readily produced at pH 7.0 under greenhouse conditions in the absence of EDTA or if Fe is omitted from the nutrient solution (Plate 115). It also appears very quickly in young seedlings transferred to Fe-free solutions and in older plants from which Fe is withheld. The younger leaves change from green to light green and finally to a uniform light yellow (Plate 116). The veins remain green at first, but later they too become bleached. As the symptoms progress in new leaves, older leaves are also gradually affected by the same pattern of chlorosis. Eventually, the bleached blades become necrotic, and the necrosis causes them to cup upward (Plate 116).

If Fe is reabsorbed before the blade tissue becomes permanently damaged, the fine veins become green and prominently netted. This symptom is normally associated with Fe deficiency but is actually a recovery symptom (Plate 117).

Iron deficiency symptoms have not been observed in commercial sugar beet fields, even on highly calcareous soils conducive to pronounced Fe deficiency symptoms in many other crops. Iron deficiency of sugar beets occurs when chlorotic blades contain less than 50 ppm of Fe.

Zinc (Zn)

A light green color on the larger leaves near the center of the plant is the first apparent symptom of Zn deficiency (Plate 118). As the chlorosis becomes more intense, small pits develop between the veins on the upper surface of the blades (Plate 119). These pits enlarge in an irregular pattern as more tissue collapses. The entire area between the veins gradually becomes dry, and the primary veins are left prominently outlined, turgid, and green (Plates 118 and 120). Finally, the blades wither, cupping upward, with the petioles remaining upright (Plate 118).

Sugar beet is not particularly susceptible to Zn deficiency under field conditions, but it has appeared occasionally in seedlings on peat soils in California. If, however, symptoms of Zn deficiency have not been observed in other crops, it is highly unlikely that sugar beet production will be adversely affected by Zn deficiency. Without local experience in correcting Zn deficiency in sugar beet, try on a small-scale Zn-containing foliage sprays or soil applications of Zn sulfate or Zn chelates as soon as symptoms have appeared. Zinc deficiency is indicated when blades contain less than 15 ppm (dry weight basis).

Manganese (Mn)

Early symptoms of Mn deficiency are not readily distinguishable from those of Cl or, at times, from Fe and Cu deficiencies. In each instance, chlorosis begins in the younger leaves (Plate 121) and can be seen as netted veining in leaves viewed against the light. As the severity of the symptoms increases, leaf blades of Mn-deficient plants gradually fade from green to a uniform yellow with a metallic grey luster. In contrast, in plants with Cl deficiency (Plate 128), the larger veins remain green and become raised; with Fe deficiency, the leaves either turn pale yellow or develop distinct green veins (Plates 116 and 117); with Cu deficiency, the veins are broad and diffuse, and the leaves become bleached and scorched (Plate 125). As Mn deficiency increases, a gray, metallic, sometimes purplish luster develops on the upper blade surface (Plate 121). This symptom is followed by gray to black freckling along the veins (Plate 122). In time, the freckles coalesce in shrunken, black necrotic areas along the larger veins (Plate 123).

As a corrective measure for Mn deficiency on calcareous soils, spray the foliage in a small-scale field trial with a recommended solution to overcome Mn deficiency. To overcome Mn toxicity on acid soils, raise the soil pH by liming the soil. Manganese deficiency is indicated when blades contain less than 20 ppm of Mn, and toxicity is indicated by similar symptoms in leaves that contain more than 5,500 ppm of Mn.

Copper (Cu)

Copper deficiency in sugar beet has not been observed in the field and is not easily produced in nutrient solutions. Apparently, only minute amounts of Cu are required for growth, since deficiency symptoms cannot be produced in nutrient solutions without using repurified salts and redistilled water. Under these conditions, plants first develop a mild chlorosis of the young, center leaves (Plates 124 and 126) similar to that caused by deficiencies of Fe, Cl, and Mn. In contrast, chlorosis caused by deficiencies of N, S, and Mo tends to occur as an overall yellowing of plants. In plants with Cu deficiency, symptoms progress from mild chlorosis to fine, green netted veining (Plate 125) and then bleaching of the blade tissue (Plate 126). This bleaching differs from the spotted necrosis of Fe deficiency, the black spotting of Mn deficiency, and the raised veining of Cl deficiency. To overcome Cu deficiency, spray the foliage in a small-scale trial with a commercially recommended Cu solution. Copper deficiency is indicated when blades contain less than 2 ppm of Cu (dry weight basis).

Chlorine (Cl)

Symptoms of Cl deficiency first appear as a chlorosis of the blades of the younger leaves near the center of the plant (Plate 127). The interveinal areas of the leaf blades become light green to yellow; the main veins remain green and become raised (Plate 128). If a leaf blade is viewed against bright light, a netted mosaic pattern can be seen branching out from the main veins. In early phases, this mosaic pattern resembles the netted veining of Mn deficiency. As Cl deficiency develops, fibrous roots become stunted, and the interveinal areas appear as flat, yellow-green depressions, which become dry and contrast sharply with the adjacent raised green veins. These advanced symptoms are unique to Cl deficiency and are clearly distinguishable from symptoms of other nutrient deficiencies. Chlorine deficiency symptoms appear when the Cl concentration in petioles taken from leaf blades with symptoms is less than 0.04% (dry weight basis). Plants appear normal with Cl concentrations in petioles up to 8.5%.

Toxicity Symptoms and Other Disorders

pH Effects

Sugar beets in acid soils are not only directly affected adversely by high hydrogen-ion activity but also indirectly by an excessive accumulation of Mn and especially of Al. In general, soil acidity has a greater adverse effect on sugar beet growth on mineral soils than on organic soils. Sugar beets grown on acid mineral soils with a pH of less than 5.0 are likely to show symptoms of Al toxicity more often than those of Mn. Blades with Mn values up to 5,500 ppm have been reported to be without toxicity symptoms; and since Mn deficiency symptoms at 20 ppm (Plates 121–123) are similar to those with true toxicity symptoms, the blades must be analyzed for Mn to tell them apart. Liming soils to approximately pH 6.0 usually prevents harmful effects from soil acidity.

Other Heavy Metals

Toxicity symptoms on sugar beets from excesses of chromium (Cr), Cu, Zn, nickel (Ni), cadmium (Cd), Mo, or lead (Pb) will resemble those for Fe deficiency. Growth is stunted, and younger leaves develop typical Fe-chlorosis symptoms. Applications of ferrous sulfate sprays are likely to decrease this young-leaf type of Fe chlorosis, but such toxicities are rarely seen under field conditions.

Boron Toxicity

Boron toxicity of sugar beet is rarely observed under field conditions. When B toxicity does occur in plants, it is chiefly affected by the concentration of B in the soil solution. Since B is minimally adsorbed by soil colloids below pH 7.0, B follows the transpiration stream, with the older leaves showing toxicity symptoms first. Boron toxicity symptoms develop as a yellow-tinted band around the leaf margins. The chlorotic zone becomes necrotic and gray, while the major portion of the leaf remains green. Older leaves eventually wilt and die. Boron toxicity in sugar beets grown in the greenhouse in solution cultures begins when respective B values exceed 250, 420, and 650 ppm (dry weight basis) in young, mature, and old blades.

Critical and Sufficient Nutrient Concentrations

Critical Concentration

The critical nutrient concentration is a convenient reference point for assessing the nutrient status of a crop, providing a "recently matured" leaf is taken as the leaf to sample for analysis. A recently matured leaf can be taken any time during the growing season, because it is approximately at the same physiological age not only for a single sampling date but for all samplings during the growing season.

The critical concentration (Fig. 7.2) can be determined hydroponically when all nutrients except the one to be calibrated are adequately supplied for growth. The calibrated nutrient is supplied, one time only, in convenient doses, starting from nil to a full treatment such as in half-strength modified Hoagland's nutrient solution. The plants are harvested when those in the first half of the series are visibly deficient and the remainder are without symptoms.

The critical concentration is taken at the point where growth is 5 or 10% below the maximum for plants that are not deficient (Fig. 7.2). Because of the greater variability of field sampling, the practical critical concentration is usually set at a higher value than that determined under controlled greenhouse conditions.

Sufficient Level

The sufficient level is the nutrient concentration maintained appreciably above the critical concentration during the growing season up to harvest for best crop production, with the exception of N. Here, a deficiency of N several weeks before harvest usually improves the quality of the crop for processing at the sugar factory.

Prognosis

Predicting the outcome of a fertilizer program based on a single plant or soil sample taken early in the growing season is fraught with inherent difficulties, primarily because of the unpredictable effects of climate on subsequent plant growth. If the climate is favorable thereafter, plant nutrient demand may often exceed soil nutrient supply and thereby induce a plant nutrient deficiency, leading possibly to a significant crop loss. Conversely, in an unfavorable climate, the soil nutrient supply may exceed demand and, in the case of N, excesses may lead to a lower quality sugar beet. In a new production area, it is better to sample fields more often initially than to sample a field only once during the growing season, particularly for the rapidly growing deep-rooted sugar beet plant. Therefore, the longer the time interval between sampling and harvest, the greater the likelihood of estimating the fertilizer needs of the crop incorrectly (Fig. 7.3).

Management Considerations

Nitrogen, the Key Element

Nitrogen is the key element in the nutrition of the sugar beet plant when all other nutrients are present in ample supply and no other factor is limiting growth from the time of planting to harvest. In this scenario, the sugar beet plant, the soil, the environment, management, and other factors are in a dynamic equilibrium from planting to harvest to produce sugar photosynthetically. More specifically, the N supply from the soil will vary almost hourly and so will the demand for N by the plant. When nitrate-N is present above the critical concentration, growth may be limited by some other nutrient or nonnutrient growth factor. When N becomes limiting, photosynthesis fortunately continues, but at a reduced rate, and sugar accumulates gradually in all parts of the plant, especially in the storage root. When the storage root is relatively small, it reaches sugar storage capacity within a few days, and when it is very large, it may take from 4 to 6 weeks. However, to make matters more complicated, nitrification in soil does not stop completely, as it does in hydroponic cultivation of sugar beets, where addition of N to the nutrient solution can be controlled. In the soil environment, nitrification of

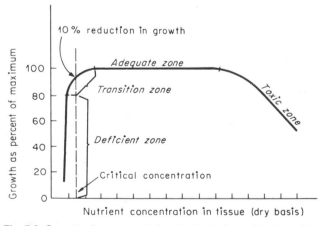

Fig. 7.2. Growth of sugar beets in relation to the nutrient concentrations of tissue of the same physiological age. The critical concentration is taken at the point where growth is 10% less than the maximum. Deficiency symptoms generally appear below the critical concentration and fail to appear above it. Symptoms may or may not appear in the transition zone. The sharper the transition zone, the more useful the calibration for diagnostic purposes. Reprinted, by permission, from Ulrich and Hills (15).

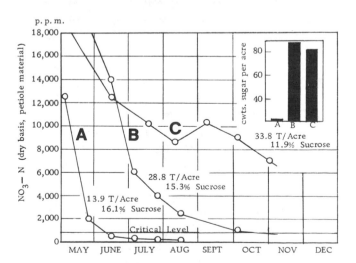

Fig. 7.3. Nitrogen concentration progression through the second-year growing season in three sugar beet fields—A, B, and C. An average application of N, such as 100 lb/acre, seldom meets the N needs of a sugar beet crop. Field A became deficient too soon according to petiole analyses, and most likely needed additional N applied. Adapted from Ulrich and Hills (15).

organic-N continues at a slow rate, and the nitrate-N formed now reacts with the surplus sugar to enhance fibrous root growth, to enlarge the storage root, and even to develop green crown leaves. When the N supply again exceeds demand, top and storage root growth renew vigorously, causing the storage root sugar concentration to decline to that of a high-N plant, and thus lowering the processing quality of the sugar beet.

Determining the soil N supply before planting and adding more N at planting to meet the calculated (estimated) yield goal, and then determining pulp nitrate-N at the tare laboratory at harvest, tells mainly the supply side of the N story. The demand side of the N story should also be told by a systematic plant analysis program. In this program, the petiole nitrate-N values will determine more fully the relationship of preplant soil N, pulp nitrate-N, beet sugar concentrations, and beet yields to sugar recovered and lost to molasses at the sugar factory. With this information, adjustments can be made in the N program for the current and succeeding crops on the same field or even on adjacent fields with the same soil type and management practices.

If it occurs just before harvest, late-season nitrification of organic N from crop residues, cover crops, and manures, and even downward movement of nitrate-N by rain, will depress sugar concentrations.

Under high residual N conditions, a sugar beet crop may serve as a deep-rooted scavenger crop to reclaim N unused by a preceeding shallow-rooted crop.

Nutrient Balance

There is no easily recognizable nutrient balance within the plant for best crop production, except when an essential nutrient becomes so low as to limit growth. When N is limiting, there is no substitute for N, as can be easily shown by growing plants hydroponically in a closed system with a single dose of N when other nutrients are in ample supply. At these times, nutrient ratios vary widely without affecting yield. Rapid growth is only restored when N is replenished.

Storage Root Size

In this scheme of events, the increase in root size is a daily sugar-using event in the presence of an ample supply of N from emergence to harvest. Unfortunately, under high-N conditions, the sugar beet plant has no internal mechanism for ripening to enhance sugar storage; only when N, the sugar-using nutrient, becomes deficient or when low night temperatures occur during autumn is sugar accumulation accelerated. When these events take place simultaneously over the 4–6 weeks before harvest, small storage roots increase in sugar concentration more rapidly than large storage roots.

Selected References

1. Carter, J. N., Jensen, M. E., and Bosma, S. M. 1971. Interpreting the rate of change in nitrate-nitrogen in sugarbeet petioles. Agron. J. 63:669-674.
2. Draycott, A. P., and Farley, R. F. 1973. Response by sugar beet to soil dressings and foliar sprays of manganese. J. Sci. Food Agric. 24:675-683.
3. Hamence, J. H., and Bram, P. A. 1964. Effects of soil and foliar applications of sodium borate to sugar beet. J. Sci. Food Agric. 15:565- 579.
4. Hepler, P. R., Struchtemeyer, R. A., and Hutchinson, F. E. 1966. Effect of lime and pH on sugar beet. Maine Farm Res. 14(2):14-16.
5. Hewitt, E. J. 1953. Metal interrelationships in plant nutrition. I. Effects of some metal toxicities and sugar beet, tomato, oat, potato, and marrowstem kale grown in sand culture. J. Exp. Bot. 4:59-64.
6. Hills, F. J., Sailsbery, R. L., and Ulrich, A. 1982. Sugarbeet fertilization. Univ. Calif. Div. Agric. Sci. Bull. 1891.
7. Keser, M., Neubauer, B. F., and Hutchinson, F. E. 1975. Influence of aluminum ions on developmental morphology of sugarbeet roots. Agron. J. 67:84-88.
8. Loomis, R. S., Ulrich, A., and Terry, N. 1971. Environmental factors. In: Advances in Sugarbeet Production: Principles and Practices. R. T. Johnson, J. T. Alexander, G. E. Rush, and G. R. Hawkes, eds. Iowa State University Press, Ames.
9. Menser, H. A. 1974. Response of sugar beet cultivars to ozone. J. Am. Soc. Sugar Beet Technol. 18:81-86.
10. Ulrich, A., and Ohki, K. 1956. Hydrogen ion effects on the early growth of sugar beet plants in culture solutions. J. Am. Soc. Sugar Beet Technol. 9:265-274.
11. Ulrich, A., Ririe, D., Hills, F. J., George, A. G., and Morse, M. D. 1959. Plant analysis—A guide for sugar beet fertilization. Calif. Agric. Exp. Stn. Bull. 766.
12. Ulrich, A., and Hills, F. J. 1969. Sugar beet nutrient deficiency symptoms: A color atlas and chemical guide. Univ. Calif. Div. Agric. Sci. Publ. 4051.
13. Ulrich, A., and Hills, F. J. 1986. Abiotic disorders. Pages 51-54 in: Compendium of Beet Diseases and Insects. E. D. Whitney and J. E. Duffus, eds. American Phytopathological Society, St. Paul, MN.
14. Ulrich, A., and Hills, F. J. 1989. Sugarbeets. In: Detecting Mineral Nutrient Deficiencies in Tropical and Temperate Crops. D. L. Plucknett and H. B. Sprague, eds. Westview Press, Boulder, CO.
15. Ulrich, A., and Hills, F. J. 1990. Plant analysis as an aid in fertilizing sugarbeet. In: Soil Testing and Plant Analysis. 3rd ed. R. L. Westerman, ed. Soil Sci. Soc. Am. Book Ser. 3.

Chapter 8

J. B. Sinclair
Department of Plant Pathology
University of Illinois at Urbana-Champaign

Soybeans

Soybean (*Glycine max* (L.) Merr.) plants produce erect stems that are stiff and hairy, with some side shoots, which may be more or less vinelike. Compound leaves have three leaflets, each up to 6-in. (15-cm) long and 4-in. (10-cm) wide, borne on long petioles. Lateral roots are formed soon after the radicle of the seed begins to elongate. The taproot of soybean plants is less pronounced than those of some other legumes, such as alfalfa. Soybean taproots branch and rebranch, and within 5 to 6 weeks after planting they will reach the center of a spaced row. By the end of the season, the roots penetrate to a depth of 3–6 ft (0.9–1.8 m) in well-drained, loose soil. However, the bulk of the roots grows in the top 1 ft (30 cm) of soil, with extensive growth in the upper 6 in. (15 cm).

Soybean plants have a typical legume flower, which is inconspicuous, white or purple, with a hairy calyx. The upper two petals are more or less united, and the standard of the corolla is broad. Stamens usually are fastened together. The fruit is a pod that hangs on a short stalk or pedicel, up to 2.5–3 in. (6–8 cm) long and 0.75-in. (2-cm) wide. At maturity, pods are brown and hairy, with two to four globular seeds that vary in color depending on the cultivar or type.

Most plants have two growth phases, a vegetative and a reproductive or flowering stage. A standard method for reporting soybean growth stages has been devised and should be used in reporting information associated with growth stages (details are given in Sinclair and Backman [3]). The period between emergence and appearance of the first flower is the vegetative period, and the period from the first flower to the death or maturity of the plant is the reproductive stage.

Soybean plants are photoperiod sensitive. They make the transition from vegetative to reproductive stages in response to day length, or more accurately, to the length of darkness in a 24-hr period. This photosensitivity should be considered in choosing cultivars for the region or area of production.

The height attained by soybean plants before flowering depends on environmental conditions. The amount of growth after the initiation of flowering depends on growth habit as well as on environmental conditions: some cultivars are determinate and some are indeterminate. The height of determinate cultivars increases very little after flowering. Indeterminate cultivars grow two to four times higher after flowering begins. As with day-length sensitivity, growth habit needs to be considered with regard to the region or area of production.

In many locations, nitrogen-fixing bacteria (*Bradyrhizobium japonicum* (Kirchner) Jordan) are introduced when seeds are planted. Inoculation is necessary where soybeans have not been grown before, where the interval between soybean crops has been many years, or if environmental conditions are not favorable for the overseasoning of the bacterium. *B. japonicum* takes gaseous nitrogen from the atmosphere and fixes it into forms easily used by soybean plants, thus reducing or eliminating the use of commercial nitrogen fertilizers.

Soybeans require relatively large amounts of phosphorus and potassium and generally respond well to fertilizer. Soil test readings for soil pH, phosphorus, and potassium that are favorable for corn (maize) are also ideal for soybeans.

In most locations where soybeans are grown, the amount of water used by the growing crop plus that lost by evaporation from the soil exceeds the normal rainfall during the growing season. Therefore, water must be recovered from that which is stored in the soil or from irrigation. On most soils and during most growing seasons, there is sufficient moisture to meet plant requirements.

Deficiency Symptoms

Nitrogen (N)

Nitrogen-deficient plants become pale green. Later, the leaves turn distinctly and uniformly yellow (Plate 129). Symptoms appear first on the basal leaves and quickly spread to upper parts. The plants eventually defoliate and often are spindly and stunted. The deficiency can be diagnosed by analyzing leaves for N, uprooting the plant to check for nodule formation by

B. japonicum, and analyzing the soil to determine its pH and Ca content.

Soybeans normally do not need N fertilization, because the requirement for this nutrient is met by the nodulating bacterium *B. japonicum* and by available soil N. The deficiency occurs when soil N is low, soil moisture is limiting, and Mo is deficient. Where soybeans are grown for the first time, seed inoculation with *B. japonicum* is required. Occasionally, N deficiency symptoms are caused by the soybean cyst nematode or other nematodes, root rot, or Mo deficiency, all of which interfere with nodule development.

Phosphorus (P)

Soybeans require relatively large amounts of P, especially at pod set. It is required for normal N fixation by *B. japonicum*.

Symptoms of P deficiency are not well defined. The chief symptom is retarded growth; affected plants are spindly and have small leaflets. Leaves turn dark green or bluish green. The leaf blade may curl up and appear pointed. Blooming and maturity are delayed. The dark green of the leaves gives the impression that the plants are quite healthy.

Diagnosis is made by leaf analysis for P content. Soil analyses for available P and soil pH also are good indicators of the nutrient availability. In acidic soils, however, P is in a form not available to soybean roots. Phosphorus uptake is also reduced in cool, wet soils.

Potassium (K)

Soybeans require large amounts of K. It is important for all aspects of plant growth and influences the plant's nutritional balance. It is also involved in the uptake of Ca and Mg. The maximum need for K uptake occurs during periods of rapid vegetative growth. Plants deficient in K tend to have weak stems and are more susceptible to some plant diseases.

Potassium deficiency symptoms are well defined. Symptoms appear first on the older leaves. In early stages of growth, an irregular yellow mottling appears around leaflet margins. The yellow areas coalesce to form a more or less continuous irregular yellow border (Plate 130). This symptom often is followed by necrosis of chlorotic areas and downward cupping of leaf margins (Plate 131). The dead tissues then drop away, so that the leaves appear ragged. Chlorosis and necrosis may spread inward to include half or more of the leaflet. The basal portions remain green. Severe deficiency tends to produce misshapen and wrinkled seeds. Injury from leafhoppers feeding on cultivars with glabrous leaves often mimics K deficiency.

Calcium (Ca)

Calcium is important for the normal development of the beneficial bacterium *B. japonicum* and for symbiotic N fixation as well as for the host plant. It affects K and Mg accumulated by plants.

Calcium deficiency is most common in acid, sandy soils low in organic matter. Plants grown in Ca-deficient soils may have N deficiency symptoms. The emergence of primary leaves is delayed, and they are cup-shaped when emerged. Terminal buds of primary leaves become necrotic, and other parts are chlorotic. Tissues between veins tend to be ridged (Plate 132). Terminal buds wither, and petioles fall. Later, the primary leaves become flaccid and drop prematurely. Plants deficient in Ca are poorly nodulated and more vulnerable to infection by damping-off microorganisms.

Magnesium (Mg)

Magnesium, an important plant nutrient, is required for chlorophyll formation, because it is part of the molecule. Magnesium deficiency also inhibits N fixation by *B. japonicum*.

In early stages of growth, Mg deficiency is recognized by pale green to yellow interveinal tissues on the leaf blade (Plate 133). Leaves later become deep yellow, except at the base. The lower leaves show symptoms first; later, rusty specks and necrotic blotches may appear between the veins and around the edges of middle and upper leaves. The deficiency gives older plants the general appearance of early maturity. Leaf margins curve down, with a gradual yellowing from the margin inward, followed by bronzing of the entire leaf. Magnesium deficiency can occur after heavy application of ammonia or K to the soil. It is most prevalent in deep, sandy, acidic soils.

Sulfur (S)

The main symptom of S deficiency is small, yellow-green leaves at the top of the plant. Stems are thin, hard, and elongated. Leaf symptoms resemble those produced by deficiencies of other elements, such as N and P, but stem elongation is characteristic of S deficiency.

The deficiency results in the accumulation of amino acids, carbohydrates, and nitrates that do not contain S. The synthesis of amino acids containing S is interrupted; thus, protein synthesis is reduced, and proteolysis is increased. Cell walls thicken from the accumulation of starch and hemicellulose.

The availability of S depends on the rate at which it is released from organic matter, which in turn is influenced by the kinds of plant residues, soil moisture, and soil pH.

Iron (Fe)

Early leaf symptoms of Fe deficiency are a yellowing of interveinal areas on younger leaves (Plate 134). Later, even the veins may become chlorotic; the whole leaf finally turns ivory to white. Brown, necrotic spots may occur near leaf margins. Plant growth is repressed.

Chlorosis caused by Fe deficiency often is referred

to as lime-induced chlorosis. Chlorotic symptoms may also appear when plants are not using available Fe effectively, and large amounts of Fe accumulate in the leaves. The deficiency often is severe in high-pH soils (usually well above pH 7.0), in which iron salts are converted to the ferric form, which is unavailable to soybean plants.

Well-aerated, alkaline soils low in organic matter, unusually low or high temperatures, and intense sunlight increase the severity of Fe deficiency. Excess Ca, Cu, Mn, Mo, P, V, or Zn induce Fe chlorosis by inhibiting root absorption of this nutrient or interfering with the utilization of it. High concentrations of Mn in poorly aerated soils may inhibit Fe absorption and utilization.

Soybean cultivars differ in their tolerance of Fe deficiency. At marginal deficiency levels, some cultivars show complete chlorosis, but others remain normal.

Zinc (Zn)

Soybean yields are considerably reduced in Zn-deficient soils. Zinc-deficient plants have stunted stems and chlorotic interveinal areas on younger leaves (Plate 135). Later, entire leaves may turn brown or gray and drop early. Few flowers are formed, and the pods that set are abnormal and slow in maturing. Yield reduction is primarily the result of the reduced number of seeds formed.

Zinc deficiency is most likely to occur in strongly weathered, coarse-textured soils that are alkaline (with free lime), in eroded soils, in soils excessively fertilized with P, and in soils low in organic matter. Adding lime and P may reduce Zn availability and thus cause a deficiency of this nutrient. Soybean cultivars vary greatly in their tolerance of Zn deficiency.

Manganese (Mn)

Soybeans are sensitive to Mn deficiency, which commonly occurs in plants grown in well-drained alkaline or neutral soils, such as muck and peat soils. Interveinal areas of leaves become light green to white; the veins remain green. The symptoms resemble the early symptoms of Fe deficiency, except the veins of Mn-deficient leaves remain green and stand out prominently (Plate 136). Necrotic, brown spots develop as the deficiency becomes more severe. Leaves may drop prematurely. The symptoms are pronounced in cool weather. When temperatures rise, affected plants generally outgrow mild deficiency symptoms. Manganese deficiency symptoms frequently appear if the Fe concentration is increased.

Copper (Cu)

Research on Cu deficiency in soybeans is limited. The deficiency reduces seed yields, but plants may not show obvious symptoms. It is likely to occur in alkaline peat and muck soils (pH 7–8), inorganic soils high in organic matter, and acid, highly leached, sandy soils. The deficiency can be detected by soil and plant analyses.

Boron (B)

Soybeans are relatively tolerant of B deficiency, which occurs in alkaline or strongly acidic soils in areas of low rainfall and in soils high or very low in organic matter. Symptoms of the deficiency include shortened internodes and yellowing or reddening of upper leaves. Flowers fail to develop, particularly in dry weather. Boron deficiency may be confused with leafhopper injury and K or Ca deficiency.

Molybdenum (Mo)

Molybdenum is required for N metabolism in leaf tissues and for N fixation by *B. japonicum*. Molybdenum deficiency occurs in highly acidic soils that are strongly weathered and leached and in soils in which the element is in an unusable form. The deficiency greatly reduces yields. Plant growth, the number of pods, the number of seeds per pod, seed size, nodulation, and the total N and protein contents of the seeds are reduced.

Symptoms of this deficiency resemble those of N deficiency and are probably caused indirectly by reduced N utilization rather than directly by lack of Mo. Leaves are pale green or yellow, necrotic, and twisted. The necrosis is confined largely to the margins, midribs, and interveinal areas.

The deficiency also can arise when seeds from plants grown in Mo-deficient soil are planted. Seeds with a low Mo content may produce plants that show deficiency symptoms when grown in soils with a marginal supply of this nutrient.

Toxicity Symptoms

Some essential nutrients are not accumulated in soybeans in sufficient quantities to be toxic. These usually are N, P, K, and Ca. Toxicity more typically occurs from micronutrients and heavy metals.

Boron

Excessive B, especially when near rows and in sandy soils, is extremely toxic to soybeans, especially seedlings. Affected plants are stunted. Toxicity symptoms appear only on leaves, usually those that are just opening (Plate 137). The pubescence at the leaf tip becomes brown. This symptom is followed by browning of the leaf tip tissues and then progressive browning of the entire leaf. The leaf continues to develop, but eventually the dead tissue at the tip and along the leaf margins prevents the blade from expanding; the leaf crinkles or puckers, and its edges cup up or down. Older leaves may fall prematurely.

Magnesium

The leaves on plants with excess Mg are dark green and appear to be vigorous (Plate 133). However, yields may be reduced.

Manganese

Manganese toxicity appears in plants growing in soils below pH 5.0 containing abundant Fe-Mn pebbles and is associated with high soil moisture and management practices that increase soil acidity. Plants are stunted, and leaves are crinkled and cup down (Plate 138). The youngest leaves are affected first. Yellowing between leaf veins may or may not be observed. The toxicity can be corrected by liming the soil to pH 6.0 or slightly higher.

Zinc

Soybeans also are sensitive to Zn toxicity, particularly in fields near industrial plants that emit large amounts of Zn or after heavy applications of Zn salts, particularly in acidic soils. Soybean cultivars vary greatly in their tolerance of Zn.

Management Considerations

A knowledge of deficiency and toxicity symptoms can be beneficially used by cooperative extension personnel, consultants, and growers to determine whether problems exist. It is often too late to correct deficiency problems after symptoms appear. If tissue analyses for N, P, Zn, and Fe are available, results allow more accurate diagnosis of specific nutrient problems.

Nitrogen

Soybeans have the ability to supply their own N needs, provided they have been inoculated with *B. japonicum* and the soil contains sufficient lime and other fertilizer elements. Often it is not necessary to supply N fertilizer except for a small amount as a starter if N fixation is delayed for about 2 weeks by low soil temperatures.

Phosphorus

Phosphorus deficiency can be corrected by application of fertilizers containing P. This is profitable when the availability of the nutrient in the soil, as determined by a soil test, is low and other nutrients are in adequate supply. The fertilizer should be well mixed into the soil because this element moves slowly through most soils. Application of P increases the number and weight of N-fixing nodules and thus increases the N-fixing capacity of the nodulating bacterium *B. japonicum*. Mycorrhizal fungi associated with soybean roots may help overcome P deficiency.

Potassium

Potassium deficiency can be corrected by applying fertilizers that contain K. Need should be determined by a soil test. Potassium improves yield by increasing the number of pods and the number, size, and weight of seeds. It also increases nodulation by *B. japonicum*, thus enhancing the N-fixing capacity of the nodulating bacteria.

Calcium

Calcium deficiency can be corrected by liming the soil to pH 6.0 or above. Calcium does not neutralize acid soil; a basic anion is required. However, most liming materials that occur in nature contain basic anion compounds.

Magnesium

Application of dolomitic limestone corrects Mg deficiency. If lime is not needed, potassium-magnesium sulfate can be used to supply Mg (as well as K and S). Application of Mg increases the oil content and yield of soybeans and improves growth.

Sulfur

Sulfur deficiency can be corrected by applying elemental S or S-containing fertilizers to the soil.

Iron

Control of Fe chlorosis is still being investigated. Since Fe shortage is due to soil fixation, application of soluble Fe salts to the soil does not correct the problem. Several sources of chelate Fe may be used. Response to foliar sprays of Fe has been variable and inconsistent. If chlorosis exists over large areas, producers should try a foliar spray containing Fe.

Zinc

Zinc deficiency can be corrected by applying a Zn salt (most commonly zinc sulfate or zinc oxide) to the foliage or soil. It may be broadcast and worked into the soil or placed in a band at least 2 in. (5 cm) to the side and 2 in. (5 cm) below the seeds. A single broadcast application generally is enough for 2-4 years. Seed treatment with Zn salts also is helpful. The toxicity can be corrected by liming soil to pH 6.2-6.5.

Manganese

Manganese deficiency can be treated easily with fertilizers that contain Mn. Aerial application of this nutrient to young deficient plants can increase soybean yields. The element is best applied as a spray of Mn sulfate when plants are 4-6 in. (10-15 cm) tall; one application usually is sufficient. Another good way to correct the deficiency, especially recommended for soils high in organic matter, is to mix Mn salts with acid-forming fertilizers, such as superphosphate, and apply the mixture in bands beneath the rows. The availability of Mn is effectively increased by combining it with N, P, and/or K. Manganese must be applied yearly, because alkaline soils bind it in an unavailable form.

Copper

Copper deficiency can be corrected by applying copper sulfate or chelate Cu compounds to the soil or by changing the soil pH to slightly acid (pH 6.0-6.5).

Boron

Boron deficiency is rare on soybeans in the field. The element is least available in strongly acid and alkaline soils. Adjusting soil pH for maximum growth of soybeans usually will provide enough B for normal growth. The range between B deficiency and toxicity is narrower than for any other micronutrient. The most common fertilizer sources for B are the sodium borates and boric acid. Care should be taken to avoid excess, since toxicities frequently occur.

Molybdenum

Molybdenum deficiency can be corrected by treating seeds with sodium molybdate or ammonium molybdate. Foliar sprays with these salts at the problem stage also are effective. In many soils, the deficiency can be remedied by liming the soil to pH 6.0–6.3.

Selected References

1. Rapper, C. C., Jr., and Kramer, P. J. 1987. Stress physiology. Pages 590-641 in: Soybeans: Improvement, Production, and Uses. 2nd ed. J. R. Wilcox, ed. American Society of Agronomy, Madison, WI.
2. Scott, W. O., and Aldrich, S. R. 1983. Modern Soybean Production. S & A Publications, Champaign, IL.
3. Sinclair, J. B., and Backman, P. A. 1989. Compendium of Soybean Diseases. 3rd ed. American Phytopathological Society, St. Paul, MN.
4. Sprague, H. E., ed. 1964. Hunger Signs in Crops: A Symposium. 3rd ed. David McKay, New York.

Chapter 9

D. H. Smith
Department of Plant Pathology and Microbiology
Texas A&M University
Texas Agricultural Experiment Station, Yoakum

M. A. Wells
Department of Plant Pathology and Microbiology
Texas A&M University
Texas Agricultural Experiment Station, Yoakum

D. M. Porter
USDA-ARS, Tidewater Agricultural Experiment Station
Suffolk, Virginia

F. R. Cox
Department of Soil Science
North Carolina State University, Raleigh

Peanuts

The peanut plant is a self-pollinating, annual, herbaceous legume. It is unusual because it develops flowers above ground and pods containing edible seeds below ground. Seeds are eaten raw or cooked. They are nutritious, containing 25% protein and no cholesterol. They may be boiled, broiled, roasted, fried, ground into peanut butter, or crushed for oil.

The peanut is indigenous to South America. It is found in the wild state growing erect or prostrate. The cultivated peanut (*Arachis hypogaea* L.) is characterized by growth habit. Bunch types grow upright, covering the soil with a dense foliage canopy. Plants have a well-developed taproot with many lateral roots. Roots are usually devoid of hairs and lack a distinct epidermis. Nutrient absorption takes place 3–4 in. (8–10 cm) behind the root cap.

Peanut leaves are alternately arranged in a spiral. They are pinnate, with two opposite pairs of leaflets 1–2 in. (2–5 cm) long. The petioles are 1–3 in. (2–7 cm) long. Flowers are borne on inflorescences resembling spikes, which are located in the leaf nodes. Nodes have from one to many flowers. The hypanthium is filiform. The calyx has two lobes. One is awl-like and located opposite the keel. The other is broad and located opposite the back of the standard. The five petals consist of a yellow-to-orange standard, two yellow-to-orange wings, and two petals fused to form an almost colorless keel. The flower has 10 monadelphous stamens, two of which are sterile. The pistil has an ovary containing a single carpel with one to five ovules and a long filiform style that elongates along the hypanthium and ends in a featherlike stigma among the anthers. The first flowers appear 3–6 weeks after planting, but the maximum number of flowers is produced 6–10 weeks after planting. Anthesis and pollination take place at sunrise. Self-pollination occurs in the closed keel of the flower. The flower withers 5–6 hr after opening, after which the hypanthium is shed, leaving only the fertilized ovary.

Within about 1 week after fertilization, a pointed, needlelike structure commonly called the "peg" (the gynophore) develops and elongates quickly. The fertilized ovaries are located behind the tip of the peg. The cells at the tip of the ovary become lignified and serve as a protective cap as the peg enters the soil. The peg (positively geotropic) grows into the soil to a depth of 1–3 in. (2–7 cm). With a loss of geotropism and growth of peg meristem due to a light-sensitive reaction, the tip orients itself horizontally. Then the ovary enlarges rapidly and pod growth begins.

The mature pod (fruit) is an indehiscent legume containing one to five seeds. The endocarp consists of parenchymatous tissue surrounding the developing seed. It collapses as the pod matures, forming a thin, papery lining. Mature seeds are cylindrical or ovoid. Seeds have two large cotyledons, an epicotyl, a hypcotyl, and a primary root.

Variations in *A. hypogaea* are of two main structural types (subspecies) that differ in branching patterns and seed dormancy. Each subspecies is further subdivided into two botanical varieties that correspond to market types. Forms in the subspecies *hypogaea* have alternate branching, a spreading or bunching habit, and a longer maturation period. Their seeds undergo a period of dormancy after maturity. Botanical variety *hypogaea* corresponds to the Virginia and Runner types, and botanical variety *hirsuta* is the Peruvian humpback or Chinese dragon type. Forms in the subspecies *fastigiata* have sequential branching and an erect growth habit, mature in a relatively short time, and have little or no seed dormancy. Botanical variety *fastigiata* corresponds to the Valencia market type and botanical variety *vulgaris* to the Spanish market type.

Plant analysis is an excellent method to assist in solving problems of poor growth and to monitor the effectiveness of a fertilizer program. For peanuts, the most recently matured leaf is usually sampled to obtain, as much as possible, a common physiological age. For certain nutrients, notably N, P, and perhaps Zn, concentrations decrease during the season, so selection of a common sampling time is important. The stage of growth used most commonly is just before flowering. Considering this, a list of the nutrients and estimated critical levels in leaves are given in Table 9.1. Estimates of nutrient removal from soil by the peanut crop are also available. Estimated amounts of nutrients removed are given in Table 9.2, based on a pod yield of 3,000 lb/acre (3,360 kg/ha).

Deficiency and Toxicity Symptoms

Nitrogen (N)

Nitrogen occurs in most cellular components of the peanut plant. It is a constituent of the chlorophyll molecule and plays an important part in the synthesis of proteins. The supply of fixed N frequently limits plant growth. The peanut plant is a highly efficient N-fixing legume, provided that the proper strains of bacteria (*Bradyrhizobium* spp.) are present in the soil. Bacteria penetrate the cellular tissue of the fibrous root system. This penetration process stimulates the cortical cells of the root to begin dividing and producing root nodules.

The establishment of a relationship between the bacteria and root cells initiates the N-fixation process whereby N is made available to the entire plant. Much of this N is combined with carbohydrate derivatives to form amino acids. For proper growth, small quantities of N are needed by young seedlings until N is made available through the N-fixing process.

Nitrogen deficiency in peanut plants is characterized by varying degrees of foliar chlorosis (Plate 139). Young plants not yet adequately colonized by bradyrhizobia usually appear lighter green than normal (Plate 140). In severe cases, the entire leaf becomes a uniform, pale yellow, and stems may be slender and elongated. As the plant develops, lower older leaves are most affected because soluble N from older leaves moves to new younger leaves. Stems may appear reddish because of accumulation of anthocyanin pigments.

Nitrogen deficiency chlorosis can result from lack of nodulation associated with an inadequate amount of soil bacteria, or from insufficient N reduction attributable to Mo deficiency, a condition usually associated with extreme soil acidity. Chlorosis may also result from translocation of a limited supply of N to developing fruit late in the season, or from water-logged conditions that limit root respiration and inhibit N fixation. Symptoms often appear within 2 weeks of planting.

Plants with excessive amounts of N usually have dark green leaves, an abundance of foliage, and a poorly developed root system. In peanuts, yield response to applied N varies from none to increases directly related to N fertilization. However, high rates of N fertilization can reduce pod yield.

Phosphorus (P)

With the exception of N, P is often the most limiting element in soils. Numerous P-containing organic compounds, such as lecithins, are components of biological membranes. Other compounds containing P are in new genetic material involved in energy transfer. Materials such as nucleotide triphosphates are involved in the

Table 9.1. Critical levels of plant nutrients in peanuts[a]

Nutrient	Leaf critical level
N, %	3.5
P, %	0.25
K, %	1.5
Ca, %	0.5
Mg, %	0.25
S, %	0.15
Fe, ppm	50
Zn, ppm	20
Mn, ppm	20
Cu, ppm	4
B, ppm	20
Mo, ppm	<1

[a] Reprinted, by permission, from Cox and Perry (2).

Table 9.2. Estimated amounts of nutrient removal from soil[a]

| Nutrient | Amount[a] | |
	lb/acre	kg/ha
N	107	120
P	6	7
K	22	25
Ca	4	4.5
Mg	4	4.5
S	6	7
Fe	0.26	0.3
Zn	0.09	0.1
Mn	0.07	0.08
Cu	0.04	0.05
B	0.05	0.06

[a] Based on 3,000 lb/acre (3,360 kg/ha) yield.

synthesis of RNA, DNA, the nucleus, and cytoplasm. Phosphorus is also used to make adenosine triphosphate, which is the energy source for many of the processes in cells, such as photosynthesis.

Upon entry into the root, P is rapidly converted into organic forms that are readily translocated through the plant. When the availability of P is limited, P is transferred from older to younger plant tissues. As the plant matures, most of the P within the plant is translocated into the seed and fruit.

Deficiency of P results in poor growth, because the availability of carbohydrates to the cells is limited. Leaf size is markedly reduced, and plant maturity is often delayed. Affected leaves may first become a bluish green; however, as leaves mature, they become thickened and leathery and characterized by a dull, dark green color. In time, the older leaves turn orange-yellow, and veins become reddish brown from the accumulation of anthocyanin pigments. Accumulation may also occur in the stems. Eventually, the entire leaf becomes brown and finally drops. Deficiency of P also reduces flower production and fertilization. Pod yields are reduced accordingly. Symptoms usually appear within 4 weeks after planting.

Inadequate P is probably the most common nutrient deficiency of peanut, but it can be readily corrected by application of a phosphate-containing fertilizer. Since a very small amount of P is removed by peanut plants, and essentially no P leaches, fertilization requirements are usually minimal. Excessive fertilization with P is expensive and reduces availability or utilization of micronutrients. In a study comparing P deficiency of two cultivars, Early Runner and Florigiant, stem symptoms differed. In Early Runner, the main axis of each branch split longitudinally to the node where the last leaf was still intact, and the stem gradually dried. A reddish brown color developed in Florigiant stems.

Potassium (K)

Potassium is an osmotically active element that regulates turgidity of cells. Potassium is an essential element involved in the process of opening and closing of stomatal pores by diffusing into and out of the guard cells. It is also involved in the translocation of photosynthates, and it activates numerous enzyme systems. By activating enzyme systems more efficiently than other cations, it affects carbohydrate formation, breakdown and translocation of starch, nitrogen metabolism, and protein synthesis. Potassium is easily redistributed from mature to younger organs; therefore, deficiency symptoms are first observable in the older, lower leaves.

Potassium deficiency is expressed by chlorosis of the leaves, beginning at the leaf margin. Some chlorosis may be interveinal, but most yellowing occurs at the leaf margins (Plate 141). These margins change to reddish brown and then become necrotic or "scorched." Severely affected older leaves are shed. Leaf margins sometimes curl upward, and some tissue near the tips of branches may redden and die. Potassium deficiency decreases the translocation of amino acids, organic acids, and sugars to roots. Potassium deficiency occurs frequently in acidic soils, and symptoms usually appear within 5 weeks of planting.

The effect of K fertilizer on peanut pod yield is variable. Peanut roots are highly efficient in obtaining residual soil K when peanuts are grown in rotation with other fertilized crops, e.g., corn, and it is not generally necessary to fertilize with K when they are planted in such rotations. If needed, moderate applications of K should be incorporated into the soil prior to planting.

Calcium (Ca)

The main function of calcium is the incorporation of this element into the middle lamella of the cell wall, thereby causing the semifluid structure to stiffen. As a component of the cell wall, Ca is required for cell elongation and division. In addition, Ca also regulates the differential permeability of cell membranes. When plants are grown without adequate Ca, the cell membranes become "leaky" and lose their effectiveness as barriers to the free diffusion of ions.

Calcium in the xylem sap is translocated upward in the transport system, but once deposited, it is almost immobile. As a result of this immobility, deficiency symptoms are always most pronounced in young tissues—meristematic zones of roots, stems, and leaves—where cell division is occurring. Without a supply of Ca, deficiency symptoms usually appear within a week. Most soils contain sufficient amounts of Ca for plant growth, with the exception of acidic soils where high rainfall occurs.

The peanut is especially sensitive to Ca deficiency. A deficiency of this element is expressed not only quickly but also with marked effects. Roots are severely affected, becoming short, stubby, and discolored. Young leaves soon wilt, and apical buds die. Stem elongation ceases, petiole regions break down, and plants are stunted. In less severe cases, leaves are small, and plants appear bushy. Brown spots or pitted areas appear on leaves, causing them to appear bronze. Flowering and fruiting are inhibited.

In the field, Ca deficiency is more commonly expressed by abnormal fruiting (Plate 142). Calcium is not translocated in the phloem to the fruit; it is absorbed by the developing pegs, and extreme deficiency of Ca in the fruiting zone completely inhibits pod formation. With a less extreme deficiency, the seeds often abort, and only the shells develop, resulting in empty pods that are called "pops." Seeds that do develop often have a darkened plumule (Plate 143). The viability of seeds, including those that do not have darkened plumules, is directly related to their Ca concentration. Cultivars vary in sensitivity to a lack of Ca in the fruiting zone. Spanish market type cultivars do not require as much Ca in the fruiting zone as the large-seeded Virginia market types.

There are two readily available sources of Ca that can be used for peanuts, agricultural limestone (calcium carbonate) and gypsum (calcium sulfate). In acid soils, agricultural limestone is used to correct soil acidity and to supply Ca. Gypsum is applied when soil pH is in the correct range, but additional Ca is needed in the fruiting zone. When required, gypsum is applied over the row at early flowering stage.

Magnesium (Mg)

Magnesium is a component of chlorophyll and is necessary for amino acid production and fat synthesis. It is also a cofactor in almost all enzymatic activities and photophosphorylation processes that are important in plant metabolism. Magnesium is a carrier of P in connection with oil formation, and it affects seed viability.

Magnesium deficiency is exhibited as interveinal chlorosis, beginning in older leaves, then moving to younger ones. The yellowing begins at the leaflet margin and extends towards the midrib. The leaflet margin may become orange and crinkle or curl. This interveinal chlorosis occurs because the vascular bundles retain chlorophyll for longer periods than the parenchyma cells between them. Older leaflets may develop necrotic areas and then defoliate (Plate 144). Symptoms usually appear within 4 weeks after planting.

Magnesium deficiency rarely limits plant growth. Deficiency usually occurs in acidic soil and in areas of high rainfall where cations are easily leached. In acidic soils, Mg may be added in the form of dolomitic limestone. This will correct the acidity and supply both Mg and Ca. In soils that are not acidic, Mg may be added, along with K, in the form of potassium-magnesium sulfate.

Sulfur (S)

Sulfur is necessary for the formation of new cells and for chlorophyll development. Sulfur is also a major constituent of organic material, as it is a component of selected amino acids (cysteine, cystine, and methionine), which are ultimately the structural units of proteins. Sulfur is involved in oil synthesis and proteinaceous oil storage organs of the peanut. Sulfur is also important in nodule formation. It is metabolized by roots only to the extent that they require it, and most of the sulfate absorbed is translocated unchanged to the shoots. Sulfur influences both the pod yield and seed quality.

Terminal portions of the peanut plant are affected by a lack of S. In S-deficient soil, root development is restricted, and new leaflets become pale green or yellow. Leaf chlorosis caused by S deficiency resembles that of N deficiency, except that it occurs predominantly at growing points because of the greater immobility of S within the plant tissue. Plants deficient in S are stunted

but appear quite upright because of reduced branching. Sulfur deficiency causes accumulation of nitrate, amide, and carbohydrates, which in turn retards the formation of chlorophyll. Symptoms usually appear within 4 weeks after planting.

Sulfur can also be absorbed by the leaves through the stomates as a gaseous sulfur dioxide (SO_2), an environmental pollutant released primarily from burning coal and wood. Sulfur dioxide is converted to bisulfite when it reacts with water in the cells, and in this form it inhibits photosynthesis and causes chlorophyll destruction. Sulfur dioxide injury to plants is caused largely by acidic aerosols during periods of foggy weather, light rains, or periods of high relative humidity and moderate temperatures.

Peanuts require relatively low amounts of S, and many commonly used fertilizers, including gypsum, contain S. Irrigated peanuts probably receive adequate S from irrigation waters, which generally contain some sulfates. The main cause of inadequate S is the heavy use of nitrogenous fertilizers and ammonium nitrate.

Iron (Fe)

Iron is required by both the legume host and *Bradyrhizobium* for a range of physiological and biochemical processes. Of particular importance is the specific involvement of Fe in several key proteins for N fixation. Iron also serves as a catalyst in the formation of chlorophyll and acts as an oxygen carrier in metabolic processes.

Iron deficiency may develop so rapidly that intermediate or mild conditions are difficult to observe. It is characterized by a failure of chorophyll production and begins in the younger leaflets. Iron-deficient leaflets will become pale green, with a sharp distinction between the green veins and the yellow interveinal tissues (Plate 145). If Fe deficiency becomes more severe, the chlorophyll will eventually disappear from the veins, and later the leaves will become almost white. Severely affected leaves then develop brown spots, or necrosis, on the lamina. Because of the lack of Fe mobility, the younger leaflets become more chlorotic. In the field, plant growth is restricted and leaf size is reduced. Symptoms appear within a week after planting.

Iron deficiencies are found in calcareous soils. Soil treatments with fertilizers containing ferrous iron have not been effective because of rapid conversion of soluble Fe into an unavailable state. Soil applications of highly stable chelated forms of Fe applied after seed emergence are effective in reduction of Fe deficiency but are very expensive. Foliar feeding has been the conventional recommendation for alleviating Fe deficiency. Spraying with a 3% Fe solution as ferrous ammonium sulfate generally corrects Fe deficiency for a few weeks, but in severe cases three or more applications of Fe may be required. Iron deficiency can result from high levels of calcium carbonate in soil, bicarbonate in soils or in irrigation water, high levels of phosphorus leading to inactivation of Fe in plants, and soils with high levels of heavy metals such as Mg, Cu, or Zn.

Zinc (Zn)

Zinc acts as a bond between a number of enzymes and their substrates. These include several dehydrogenases, and Zn has long been known to activate carbonic anhydrase. Zinc is also involved in N metabolism; deficiencies decrease RNA, thus inhibiting protein formation. Zinc influences the auxin content in the cell through tryptophan synthesis, and it may have a role in starch formation.

Zinc deficiency results in interveinal chlorosis of recently matured leaflets. In severe cases, leaflets later turn reddish brown and defoliate; however, factors related to a retarded growth rate become the dominant deficiency symptom. Internodal length is reduced, plants are stunted, terminal growth is retarded, and new leaves develop slowly. Terminal leaflets are small, thickened, leathery, and exceptionally dark green. Zinc deficiency not only reduces pod yield but also lowers the biochemical quality of oil by decreasing the proportion of phospholipids and triacylglycerols. Soil pH has a drastic effect on Zn. As pH levels increase, the content of Zn is reduced. Another factor contributing to Zn deficiency is excess P fertilization when Zn is low in the soil. To increase Zn availability, zinc sulfate should be incorporated into the soil.

The growth of peanut plants may also be restricted by Zn toxicity. Plants acutely affected exhibit chlorosis and stunting during the seedling stage, followed by progressive tissue necrosis and finally necrosis of the entire plant. Other characteristic symptoms are purple coloration of the main stem and petioles. In severe cases, a lesion develops on the stem near the base of the plant, resulting in premature plant death (Plate 146). Zinc toxicity critical values are 12 ppm for soil and 220 ppm for plant tissue. Appearance of acute Zn toxicity occurs within 2–3 weeks after seedling emergence.

Liming reduces the concentration of Zn below acute toxic levels. Since soil pH influences Zn uptake to such a large degree for the same acid-extractable Zn value, caution should be exercised in using soil values alone to predict Zn toxicity. Immobility of Zn in soil profiles and a strong residual effect of Zn in the soil make correction of existing instances of toxic levels a problem of reducing Zn uptake by the peanut plant. Applications of Zn fertilizer to crops in rotation with peanuts or planting peanuts where large amounts of Zn have been previously applied should be avoided.

Zinc toxicities are relatively rare and tend to occur only if Zn has been applied and the soil becomes quite acidic. Zinc applications may be from prior fertilization, such as with inorganic fertilizers or sludge, or from contaminated runoff from galvanized surfaces, such as roofs and fences.

Manganese (Mn)

Manganese is an essential element that plays a structural role in the chloroplast membrane system and an important role enzymatically in photosynthesis by its involvement in oxygen evolution. It also bridges adenosine triphosphate with the enzyme complex and activates decarboxylases and dehydroxylases of the tricarboxylic acid cycle. It may regulate auxin levels by activating indoleacetic acid oxidase. At high concentrations, Mn is phytotoxic, possibly through an induced auxin deficiency.

Manganese deficiency is expressed by interveinal chlorosis in the young leaflets. Symptoms range from mild, in which leaflets are light green and regions immediately adjacent to the veins and the veins themselves remaining green, to severe, in which the entire interveinal area is chlorotic (Plate 147). After a period of interveinal chlorosis, some bronzing may occur; then older leaves develop necrotic spots and fall. Stems are slender and weak. Fruiting, and subsequently pod yields, are significantly reduced.

At low soil pH, the availability of Mn may increase to a toxic level. Toxic symptoms appear as necrotic patterns on leaflets if Mn leaf concentration reaches 4,000 ppm. Peanuts are more tolerant to high concentrations of Mn than other legumes. Liming very acidic soils to pH 5.5 decreases the solubility and uptake of Mn sufficiently to eliminate the toxicity. In cases of Mn deficiency, foliar applications are more economical than soil applications, because much lower rates of application are required. For a foliar application, a soluble source of Mn can be included in a tank mixture with fungicide, eliminating the need for an additional trip over the field.

Copper (Cu)

Copper is a vital component in the function of a number of enzyme systems involving oxidation-reduction reactions. The amount necessary to meet plant needs is the least of any of the essential elements taken up as cations. A major portion of the Cu in plants occurs in the chloroplast as a constituent of the protein plastocyanin, assisting in linking the two photochemical systems of photosynthesis. Superoxidase dismutase, another Cu-containing enzyme, is also present primarily in chloroplasts. This enzyme aids in ameliorating the effects of superoxide radicals generated in photorespiration. Cytochrome oxidase, one of many Cu-containing oxidase enzymes, functions in the respiratory process, catalyzing the transfer of electrons to O_2 from H_2O.

Under Cu-deficient conditions, young leaves are deformed and greenish yellow or chlorotic. Terminal leaflets are small, and margins curl inward, giving a cupped appearance. Yellowish white spots occur in affected regions. Necrosis develops in the tips and margins, then progresses inward until the petiole drops. The bud areas are affected, resulting in stunted plants with short branches. Pod yields are reduced.

Copper deficiency frequently occurs in organic soils, because organic matter stabilizes a considerable amount of Cu. Copper deficiency seldom occurs in mineral soils, such as those typically used for peanut production. Copper applied in fungicides increases the reserve of this nutrient in the soil. Small rates of soil-applied Cu are adequate to correct deficiencies in calcareous mineral soils. Recommendations are usually from 2 to 7 lb/acre (2 to 8 kg/ha) as $CuSO_4$ every 3 to 4 years. Soils containing 0.2 ppm diethylenetriaminepentaacetic acid (DTPA)-extractable Cu or more should not be fertilized with Cu. The chelated sources of Cu are effective and especially adapted to foliar applications. Rates of 1–4 lb/acre (1–5 kg/ha) should be used to avoid accumulation of Cu causing a possible deficiency of Fe.

Boron (B)

Experimental evidence indicates that B is essential for plants, but a vital physiological role has not been clearly defined. Because B forms complexes with sugars and related molecules, the function of B in the plant may involve long-distance sugar transport. Boron is essential for the production of pollen and seed and for the formation of cell walls. Boron is unique among the essential elements, because uptake is frequently in the form of an undissociated boric acid instead of the ionic form.

Severe B deficiency symptoms are manifested by deep green leaves. Water-soaked areas develop, and lesions occur on leaves, petioles, and stems. Growth of the plant is restricted. Terminal leaflets become small and deformed; internode length is reduced, and secondary branching occurs, resulting in a short and stumpy plant. The lower branches split. Flowering and pod production are reduced. The pods of certain cultivars exhibit fine cracks. The root system is stunted, subsequently turning brown and brittle. Activity of the lateral root meristems is reduced, and the tips become decayed. Lateral roots appear slightly shorter and thicker.

The most common B deficiency symptom in field-grown peanut plants occurs in the fruit. The seeds do not develop properly. The inner face of the cotyledons is depressed in the center and that region turns brown, especially when the seeds are roasted (Plate 148). The symptom is known as "hollow heart." Boron deficiency occurs in highly weathered sandy soils. Critical B deficiency levels are below 0.2 ppm for calcareous soils and 0.05 ppm for acidic soils.

Whenever B is applied, only small quantities are necessary for optimum production. Boron toxicity symptoms are likely to occur when the quantity exceeds 7 oz/acre (0.5 kg/ha). Leaflet margins become chlorotic, necrotic, and in severe toxicity the leaflets defoliate. Boron toxicity also causes antagonistic action on Cu. The activities of two Cu protein enzymes decrease in the leaves of plants when B occurs at a toxic level.

Molybdenum (Mo)

Molybdenum activates the nitrate reductase enzyme in which nitrate is reduced to nitrite, beginning the reduction of N to ammonium for amino acid synthesis. In legumes such as peanuts, Mo serves a second vital and unique function. It is a component of nitrogenase, an essential enzyme necessary for N-fixing plants. Molybdenum is required only in root nodules.

Because the primary role of Mo is in N transformations, Mo deficiency is similar to N deficiency (Plate 149). Availability of Mo, unlike that of most other micronutrients, increases with an increase in soil pH. Liming acid soils raises the pH and increases Mo availability. The amount of Mo required by plants is very low, so caution should be used when it is applied. Plants can accumulate much larger concentrations than needed for growth, but these concentrations rarely occur under field conditions.

Selected References

1. Cox, F. R., Adams, F., and Tucker, B. B. 1982. Liming, fertilization, and mineral nutrition. Pages 139-163 in: Peanut Science and Technology. H. E. Pattee and C. T. Young, eds. American Peanut Research and Education Society, Yoakum, TX.
2. Cox, F. R., and Perry, A. 1989. Groundnut (peanut). Pages 137-144 in: Detecting Mineral Nutrient Deficiencies in Tropical and Temperate Crops. D. L. Plucknett and H. B. Sprague, eds. Westview, Boulder, CO.
3. Gopal, N. H. 1975. Physiological studies on groundnut plants with boron toxicity. II. Effect on nitrogen metabolism. Turrialba 25:144-147.
4. Keisling, T. C., Lauer, D. A., Walker, M. E., and Henning, R. J. 1977. Visual, tissue, and soil factors associated with Zn toxicity of peanuts. Agron J. 69:765-769.
5. Lachover, D., Fichman, M., and Hartzook, A. 1970. The use of iron chelate to correct chlorosis in peanuts under field conditions. Oleagineux 25(2):85-88.
6. Morris, H. D., and Pierre, W. H. 1949. Minimum concentrations of manganese necessary for injury to various legumes in culture solutions. Agron. J. 41:107-112.
7. O'Hara, G. W., Dilworth, M. J., Boonkerd, N., and Parkpian, P. 1988. Iron-deficiency specifically limits nodule development in peanut inoculated with *Bradyrhizobium* sp. New Phytol. 108:51-57.
8. Parker, M. B., and Walker, M. E. 1986. Soil pH and manganese effects on manganese nutrition of peanut. Agron J. 78:614-620.
9. Paricha, N. S., and Aulakh, M. S. 1986. Role of sulfur in the nutrition of groundnut. Fert. News, Sept., pp. 17-21.
10. Porter, D. M., Smith, D. H., and Rodriguez-Kabana, R., eds. 1984. Compendium of Peanut Diseases. American Phytopathological Society, St. Paul, MN.
11. Salisbury, F. B., and Ross, C. W. 1978. Mineral nutrition. Pages 79-92 in: Plant Physiology. J. C. Carey, ed. Wadsworth, Belmont, CA.

Chapter 10

K. G. Cassman
Department of Agronomy
University of California, Davis

Cotton

Nutrition of the cotton plant is influenced by several characteristics that distinguish the nutrient requirements of cotton from those of other field crops. First, cotton is a tropical perennial shrub that is grown as an annual crop. Indeterminate growth habit and vegetative branching provide an infinite number of potential fruiting sites unless growth is limited by low temperatures, lack of water, insufficient supply of nutrients or carbohydrates derived from photosynthesis, or other limiting factors. Second, the cotton plant has a deep taproot system with unusually low root density in the surface soil layer where available nutrient levels are greatest (12,27,29). This rooting pattern makes the cotton plant more dependent on nutrient acquisition from subsoil than most other crop plants (30,49). Diagnosis of nutrient disorders of cotton often requires evaluation of subsoil conditions, particularly where subsoils are low in potassium or calcium or toxic in aluminum (3,81). And third, unlike most annual field crops, it is the yield of lint (a cellulose fiber) rather than seed yield that determines the economic value of cotton. Deficiencies of certain nutrients reduce fiber quality as well as causing a reduction in plant growth and lint yield.

Taken together, these characteristics make diagnosis and correction of nutrient disorders in cotton a challenging proposition. In the following review, special attention will be given to the unique aspects of cotton mineral nutrition, focusing on the nutritional requirements of upland cotton (*Gossypium hirsutum* L.).

Deficiency Symptoms

The specific metabolic functions of the essential plant nutrients were discussed in Chapter 1. In the cotton plant, these nutrients also play similar metabolic roles, but the expression of foliar symptoms that result from nutrient deficiency or the toxic accumulation of certain elements differs somewhat among plant species. The distinguishing features of nutrient deficiency and toxicity symptoms in cotton will be described in relation to foliar discoloration and necrosis, leaf disfiguration, and plant morphology traits that result from nutrient disorders.

Nitrogen (N)

Early and midseason N deficiency symptoms include yellowish green leaf color, which first appears on older leaves, and reduced size of younger leaves (Plate 150). Plant height is reduced, few vegetative branches develop, fruiting branches are short, and many bolls are shed in the first 10–12 days after flowering. When N deficiency occurs later in the season on plants with a moderate load of maturing bolls, foliar symptoms appear as reddening in the middle of the canopy (similar to that shown for P deficiency in Plate 151), and few bolls are retained at late fruiting positions (45). Although an N-deficient cotton crop can be harvested earlier than one with adequate N (26,34), this results from early cutout caused by accelerated leaf senescence and a shorter flowering period. Despite this apparent "earliness," N deficiency actually delays flowering by an increase in the time to appearance of the first flower and by a greater time interval between flowering on horizontal fruiting positions on the same fruiting branch (60,80).

Phosphorus (P)

Phosphorus deficiency symptoms are not distinct. Plants are stunted, leaves become darker green than normal, flowering is delayed, and boll retention is poor (35,48). Later in the season, leaves on P-deficient plants undergo premature senescence (Plate 151). Because cottonseed contain large reserves of P, deficiency rarely occurs during early growth.

Potassium (K)

On cotton, foliar symptoms of K deficiency that occur before peak bloom are similar to K deficiency symptoms found on other broadleaf crop species. Interveinal yellowing first occurs on older leaves (Plate 152), with necrotic patches developing at leaf margins under severe

Author's present address: International Rice Research Institute, Manila, Philippines.

deficiency. In many soils, however, soil K supply is sufficient until peak bloom, when rapid dry matter accumulation in bolls begins. At this point soil supply is no longer adequate to meet the increased K demand. Late-season K deficiency results in foliar symptoms that differ from early-season deficiency (77).

After peak bloom, K deficiency symptoms first appear on the younger mature leaves in the upper third of the canopy (Plate 153). Symptoms begin as slight interveinal yellowing that rapidly changes to a bronze-orange coloration (Plate 154). Leaves curl downward and become thickened, and necrotic patches may occur at the margins (Plate 155). As late-season K deficiency becomes severe, boll retention at later fruiting positions decreases markedly (15), and premature defoliation may occur (9). These symptoms are so devastating that symptoms of late-season K deficiency are sometimes mistakenly attributed to a plant disease, particularly Verticillium wilt (*Verticillium dahliae* Kleb.). The two symptoms are distinct, however, upon careful examination. Verticillium wilt causes necrotic lesions between leaf veins that have well-defined borders and are a rich brown color (Plate 156). In contrast, late-season K deficiency produces a bronze-colored leaf, and if necrotic patches develop, they occur along leaf margins without distinct borders (Plate 155). Since a plant can be both K deficient and infected by a pathogen, petiole K analysis at early bloom before foliar expression of late-season K deficiency or disease symptoms provides a useful diagnostic tool.

Calcium (Ca)

Calcium deficiency primarily affects root growth. Root tips turn brownish, root extension is inhibited, lateral branching is reduced, and taproots are small in diameter (42). Where subsoils are Ca-deficient, cotton plants often show symptoms of water stress during brief dry periods resulting from lack of root development in the subsoil where moisture reserves are adequate. Under severe Ca deficiency, which has only been demonstrated in sand culture, root tips die, yellowing and necrotic lesions occur on the cotyledons, and the terminal shoot meristem dies (86).

Magnesium (Mg)

In the field, symptoms of Mg deficiency first appear on older leaves as a distinct reddish purple interveinal discoloration (Plate 157). In the greenhouse, however, Mg deficiency occurs as interveinal chlorosis without purple color (63). It has been suggested (40) that this discrepancy results from removal of ultraviolet light by glass used in greenhouses, but this hypothesis has not been tested.

Sulfur (S)

Unlike N, S is not mobile in the cotton plant, and thus S deficiency symptoms occur on younger leaves in the upper canopy, while older leaves retain a normal green color (Plate 158). Sulfur-deficient leaves turn pale green, then a yellowish green similar to N-deficient leaves, but leaf veins tend to remain somewhat greener than interveinal tissue (Plate 158). Plants deficient in S are short; they have few vegetative branches and small bolls.

Manganese (Mn), Zinc (Zn), and Iron (Fe)

Leaf cupping and interveinal chlorosis on younger leaves indicate a deficiency of Mn, but these symptoms have only been demonstrated in solution culture. A deficiency of Mn delays the appearance of the first flower (46). Zinc deficiency also appears on younger leaves as interveinal yellowing with distinct green veins (Plate 159), and Fe deficiency symptoms are reported to resemble Zn deficiency (13). Because symptoms are similar, distinguishing among Mn, Zn, and Fe deficiencies requires soil and plant testing to identify the most likely limiting nutrient. Foliar sprays of each micronutrient separately can also be used to identify which of the three promotes a return to normal leaf color after application.

Boron (B)

Symptoms of B deficiency are distinct, and descriptions of this disorder are quite similar across a range of field environments (61,74). Petioles of younger leaves are short and thick, with dark concentric bands along their length. Flowers also are distorted in shape. With acute B deficiency, the terminal meristem may die or become deformed, the pith inside petioles may turn brown, and leaves may be deformed. Because B is required for normal floral development, flower and boll shedding increase as the severity of B deficiency increases (41). Rank growth sometimes occurs when plants are only moderately B deficient because few bolls are set.

Other Micronutrients

Deficiencies of the other essential plant micronutrients, including copper (Cu), chloride (Cl), and molybdenum (Mo) are not known to occur in cotton grown under field conditions. Symptoms caused by Cu or Cl deficiency have not been definitively characterized. In solution culture, Mo deficiency was expressed on leaves of fruiting branches as interveinal chlorosis (5).

Toxicity Symptoms

Aluminum (Al)

Phytotoxicity results from the accumulation of several elements to levels in tissue that cause toxicity in cotton. Under field conditions, Al and Mn toxicities are the problems of greatest concern. The activity of Al and Mn in the soil solution increase as soil pH decreases (i.e., more acid), and lime application to raise soil pH is a corrective measure for toxicities of these elements. In the southeastern United States, soils are often acid, and Al or Mn toxicity in cotton occurs on unlimed soil (1,2,4). Aluminum binds to the cell wall such that cell growth is restricted and interferes with the uptake, transport, and metabolism of other nutrients, particularly P and Ca. The severity of Al toxicity, however, is not correlated with tissue concentrations of Al or P in roots or shoots (25). Symptoms of Al toxicity, like Ca deficiency, first appear on roots that become stubby, swollen, and gnarled (73). In rainfed systems, Al toxicity problems are sometimes indicated by plant water stress during relatively short-duration rainless periods, because root development in acid subsurface soil is limited by high levels of Al.

Manganese

Manganese accumulates in leaves. At concentrations above 2,000 ppm of Mn in leaf blades, toxic symptoms are expressed as leaf crinkling (Plate 160).

Boron

Toxicity from excess B has been demonstrated in solution culture, but the only report of B toxicity under field conditions resulted from use of K fertilizers contaminated with relatively large quantities of borax (76). Symptoms of B toxicity develop as yellowing on leaf margins and in patches between leaf veins that become necrotic. Such necrotic areas have been found to contain 2,700–6,400 ppm of B (66), indicating localized enrichment at toxic levels.

Soil Salinity

Soil salinity affects a large number of soil properties and leads to a reduction in soil quality, which can have a negative impact on cotton production. For example, stand establishment is often poor on saline soil because of a hard surface crust that impedes seedling emergence. Once the stand is established, the cotton plant is relatively tolerant of soil salinity compared with other crops. In irrigated systems, however, soil salinity levels may build up in portions of the field that are poorly drained, or when irrigation water is of poor quality. Cotton growth and yields can be markedly lower in such areas.

Although the negative effects of salinity are not limited to nutrient disorders, soil salinity is most often associated with accumulation of sodium (Na), chloride (Cl), and sulfate. Excessive Na uptake results in toxicity symptoms that first appear on older leaves as reddish spots between leaf veins that develop into reddening of the entire leaf (56). Although the essentiality of Na in small quantities has not been demonstrated for cotton (70), Na can partially offset reductions in growth and yield caused by K deficiency (57). Sodium does not, however, replace the K requirement for fiber elongation and secondary wall thickening (16,20). When Na levels are high relative to the availability of K, uptake of K is reduced (19,72,75). Excessive Cl or sulfate in solution culture also has been found to reduce cotton growth, but foliar symptoms

were not distinct other than a general reduction in plant size (22).

Under field conditions, plant symptoms of salinity damage appear to integrate a number of stresses as discussed above. Leaves may turn reddish orange, and under severe stress, necrotic zones occur on leaf margins. Plants retain few bolls and are short with few vegetative branches. Sometimes Zn or Fe deficiency symptoms are induced on saline soils where alkalinity reduces the availability of these micronutrients. Diagnosis of a soil salinity problem is best made by measuring the electrical conductivity of a saturated soil extract (58).

Nitrogen and Phosphorus

Excessive supply of nutrients such as N or P may reduce cotton yield, although not from a toxic response to the nutrient itself. A large oversupply of N promotes vegetative growth and reduces early boll retention, causing rank growth. At harvesttime, a rank cotton crop is difficult to defoliate because of high leaf N concentration and vigorous growth of the foliage. Excessive P application to soil reduces the solubility of Zn and may induce a Zn deficiency. Oversupply problems with nontoxic nutrients are avoided by using appropriate rates of fertilizer application based on soil testing and in-season plant tissue testing.

Nutrient Requirements

Over the long term, adequate plant nutrient supply is achieved by maintaining a balance between nutrient inputs to and outputs from the soil. The immediate nutrient balance for one growing season depends on plant demand (determined by dry matter accumulation and yield potential), the nutrient supply from soil and applied fertilizer or organic nutrient sources, and nutrient losses from leaching, erosion, or other loss path-

Table 10.1. Proportion of total plant nutrient accumulation removed with harvested seed cotton

Lint yield level (lb/acre)	Proportion (%) of total aboveground plant nutrient in seed cotton			
	N	P	K	Reference
500	53	66	30	21
1,250	51	55	20	7
1,490[a]	60	52	29	33
1,520	46	46	25	31
2,420	54	63	21	...[b]
2,460[c]	55	66	28	33
Range 500–2,460	46–60	46–66	20–30	
Mean 1,610	53	58	26	

[a] Low-N fertilization.
[b] R. G. Cassman and T. A. Kerby, *unpublished.*
[c] High-N fertilization.

Table 10.2. Typical nutrient concentrations in delinted seed and lint at maturity, and nutrient accumulation and removal

| Nutrient | Nutrient concentration[a] | | Total plant nutrient uptake[b] | | Nutrient removal in seed cotton[b] | |
	Delinted seed	Lint	lb/acre	kg/ha	lb/acre	kg/ha
N	3.90	0.20	45	50	24	27
P	0.70	0.07	7	8	4	4.5
K	1.20	0.65	37	41	10	11
Ca	0.20	0.04	20–45	22–50	1	1.1
Mg	0.60	0.06	8–20	9–22	4	4.5
S	...	...	6–12	7–13	...	...

[a]Seed cotton is assumed to be 35% lint (i.e., gin turnout) and 65% fuzzy seed after ginning but only 60% by weight as delinted seed after acid-delinting.
[b]Based on 1,000 lb/acre (1,120 kg/ha) seed cotton yield for cotton grown with adequate nutrients.

ways. Seed cotton yields vary considerably, however, because of difficulties in stand establishment, boll shedding caused by insect damage or temporary water stress, disease pressure, and seasonal heat unit accumulation, all of which differ from year to year. This yield variability makes it difficult to predict the total crop nutrient demand.

Although seed cotton yields may fluctuate, the proportion of total plant N, P, and K that is partitioned to seed and lint remains relatively constant over a wide range of yields and environmental conditions (Table 10.1). In the referenced field studies, lint yields ranged from 500 to 2,460 lb/acre (560 to 2,750 kg/ha), yet the mean proportion of total nutrients in aboveground biomass found in seed cotton at maturity was 53% (±5% standard deviation [SD]) for N, 58% (±8% SD) for P, and 26% (±4% SD) for K. Because the nutrient concentration in seed and lint is also relatively constant, a reasonable estimate of the total crop nutrient requirement can be made on the basis of 1) the mean proportion of total plant N, P, or K that is partitioned to seed cotton as shown in Table 10.1, 2) the typical nutrient concentration of seed and lint (7,16,32,35,53), and 3) the expected seed cotton yield (Table 10.2). By these calculations, a 3,000 lb/acre (3,360 kg/ha) seed cotton crop would accumulate 135 lb of N/acre (151 kg/ha), 21 lb of P/acre (24 kg/ha), and 111 lb of K/acre (124 kg/ha) in aboveground biomass.

Cotton bolls, including the bur (or carpel wall), seed, and lint, are the dominant sink for N, P, and K within the plant. Maximum demand occurs after early bloom as rapid boll growth begins. The rate of uptake required to meet the demand in developing bolls depends on the boll load. At medium to high yield levels, peak uptake rates are estimated to range from 2 to 6 lb of N/acre-day (2.2 to 6.7 kg/ha-day), 0.3 to 0.7 lb of P/acre-day (0.3 to 0.8 kg/ha-day), and 3 to 5 lb of K/acre-day (3.3 to 5.6 kg/ha-day). It is noteworthy that peak N and K uptake rates are usually equivalent. Whereas N as nitrate moves readily to the root with the mass flow of water, K is less mobile in soil and moves primarily via diffusion. Root density is the most important factor governing K uptake from soil, yet the cotton root system has lower root length density than other field crops. Poor root development in surface soil may contribute to the greater sensitivity of cotton to K deficiency noted in field comparisons of several crop species (17).

When the supply of N, P, or K becomes limiting, relatively more of these nutrients are partitioned to developing bolls at the expense of nutrient supply to leaves and stems. The dominance of fruiting structures as a sink for N is reflected by the increased proportion of N partitioned to seed cotton in plants grown under N-deficient conditions shown in Table 10.1 (33,36). Similarly, the proportion of total plant K in fruiting structures (including burs and immature bolls) increased from 70% with an adequate soil K supply to 90% on severely K-deficient soil (15). Under severe deficiencies of N or K, premature defoliation occurs, causing early cutout and poor boll retention at later fruiting positions. The strength of fruit as the dominant sink for N and K also causes the expression of N or K foliar deficiency symptoms to change markedly as the plant shifts from the vegetative to reproductive growth phase.

Total Ca and Mg requirements of cotton are less certain, because the availability of these nutrients in most soils exceeds plant demand, and a relatively small fraction of total uptake is allocated to seed cotton (Table 10.2). Total plant Ca and Mg uptake per 1,000 lb/acre (1,120 kg/ha) of seed cotton yield varies considerably among soils (7), but the concentration of Ca and Mg in seed and lint is more consistent. This consistency allows a reasonable estimate of Ca and Mg removal with harvested seed cotton. Although S deficiency and a yield response to applied S are relatively common in the southern United States and in several African countries (35,77), S content of seed and lint have not been reported. Reports of total S uptake by cotton indicate considerable variability.

The total quantity of micronutrients required to achieve high lint yield is very small, and thus removal of these nutrients in seed cotton has received little attention. Correction of micronutrient deficiencies only requires a small input of the limiting element, often on the order of 0.5–5 lb of element per acre (0.6–5.6 kg/ha). Research on micronutrient requirements of cotton has focused on identifying critical tissue nutrient concentrations and adequate soil test levels for diagnosis of deficiencies or toxicities of these elements.

As for all plants, a deficient supply of any essential plant nutrient will reduce dry matter accumulation and yield of cotton. A specific requirement for K and B, however, has been demonstrated for the process of fiber elongation using in vitro ovule culture methods (14,20). Field studies have confirmed a direct, positive relationship between K supply to the cotton plant and fiber length and secondary wall thickness at maturity, both important fiber quality traits (9,16). Limited evidence suggests that B and/or Mn deficiency may also reduce fiber quality (6). Deficiencies of the other essential nutrients including N, P, and S are not thought to influence fiber quality.

Critical and Sufficient Tissue Nutrient Levels

Diagnosis of a plant nutrient deficiency is rarely definitive when it is based solely on foliar symptoms. This is because 1) several nutrient deficiency symptoms are similar in appearance, 2) foliar symptoms for the same nutrient deficiency sometimes differ depending on the stage of plant development, and 3) nuances of coloration and intensity of deficiency symptoms for a specific nutrient may vary on different cultivars or in different environments. Definitive diagnosis requires confirmation of visual symptoms by tissue nutrient analysis or soil testing and a by measurable growth response to application of the deficient nutrient.

Tissue nutrient tests are generally the most convenient way to confirm a suspected deficiency. The critical nutrient concentration of a specified index tissue (usually the first mature leaf petiole or blade) is defined as the concentration below which plant growth and yield are reduced. Critical tissue levels of N and K have been evaluated at several growth stages in a number of field studies (Table 10.3). These nutrients have received the most attention because soil deficiencies of N and K are widespread. For petiole nitrate levels, there is variation in the reported critical values from different studies. This suggests a critical petiole nitrate concentration "range," because the precise concentration at which N deficiency occurs may be influenced by cultivar, yield potential, and growth conditions. Critical petiole P levels have not been validated by sufficient field tests because P deficiency is not common, particularly in the cotton belt of the United States. Only estimates of adequate P concentration in cotton petioles are provided. For the three primary nutrients—N, P, and K—that are mobile within

the plant and in high demand in developing bolls, petiole levels normally decrease as the plant ages. Critical petiole concentrations of these nutrients therefore decrease at later growth stages.

Critical tissue Ca and Mg concentrations also have not been field-validated, and only estimates of adequate levels or critical values determined in greenhouse studies are available (Table 10.3). A soil test that determines Ca concentration in the soil solution appears to be a more reliable indicator of Ca deficiency (1,42). A relatively wide range has been reported for the critical S concentration of cotton leaf blades (10).

Among the required micronutrients, only deficiency of B is common. The cotton plant has a high B requirement relative to other crops (23), and B limitations often occur in coarse-textured soils that are low in organic matter. Critical leaf blade B concentrations ranging from 10 to 28 ppm have been reported from several field studies (Table 10.4). A field response to applied Mn also has been reported (6), but deficiencies of this nutrient are rare. Critical tissue concentrations for micronutrients other than B were determined in greenhouse experiments using solution culture or potted soil systems. For Fe and Cu, critical tissue concentrations have not been established. It has been suggested (82) that 85–112 ppm of Fe is the critical Fe concentration in young leaf blades, but these values are nearly twofold greater than the Fe concentration of the first mature leaf measured at numerous locations in Alabama, where Fe deficiency does not occur in cotton (18). Molybdenum deficiency was demonstrated only after growing seed obtained from cotton plants that were grown to maturity without Mo in nutrient solution (5).

Other Aspects of Cotton Nutrition

Cotton genotypes differ in their sensitivity to growth reductions caused by nutrient deficiencies or toxicities and also in the severity of expressed symptoms associated with a nutrient disorder. Greenhouse studies have demonstrated cultivar differences in tolerance of salinity (72), Mn and Al toxicity (24,25,50), Zn deficiency (71), B deficiency (38), and S deficiency (59). Under field conditions, several investigators have reported significant yield differences among cotton cultivars on K-deficient soils and differing cultivar responses to added fertilizer K (16,31,85).

A summary (43) of a number of studies reported interactions between cotton K nutrition and the incidence of several diseases. More recent evidence suggests that cultivar differences in K uptake and maintenance of leaf K supply after flowering are associated with cultivar differences in root development after peak bloom (12). Increased leaf K concentration also is associated with reduced foliar lesions caused by Alternaria leaf spot

Table 10.3. Critical and adequate concentrations for macronutrients in cotton at various growth stages

Nutrient	Growth stage	Index tissue[a]	Concentration (ppm)		Reference
			Critical[b]	Adequate[c]	
N (as NO_3)	Early square	Petiole	15,000–25,000	...	8
	First bloom	Petiole	10,000–18,000	...	35
	Peak bloom	Petiole	3,000–10,000	...	79
	First open boll	Petiole	1,000–4,000	...	...
	Late season	Petiole	...	<2,000	...
P (as PO_4)	First bloom	Petiole	...	>1,500	8
	Peak bloom	Petiole	...	>1,200	62
	First open boll	Petiole	...	>1,000	...
K	First bloom	Petiole	40,000–45,000	...	78
	50% Bloom	Leaf	18,000	...	54
	Peak bloom	Petiole	30,000	...	8
	First open boll	Petiole	15,000	...	32
	Late season	Petiole	10,000	...	...
Ca	Seasonal	Petiole	...	6,000–25,000	46
	Early bloom	Leaf	...	17,000–25,000	18
Mg	Seasonal	Petiole	...	2,000–8,000	28
	Peak bloom	Leaf blade	...	>3,000	69
S	Early square	Shoot	...	>2,000	59
	Seasonal	Leaf blade	2,400–4,000	>4,000	10

[a] Unless otherwise footnoted, all petiole and leaf index tissues are sampled from the first mature leaf, usually at the third or fourth visible node below the main stem apex. Leaf blades are sampled at the main stem node of the most recent fruiting branch to flower.

[b] All critical nutrient concentrations were derived from field studies conducted in agricultural soils. The critical concentration is the level below which foliar deficiency symptoms are likely to develop, and a yield response to the applied nutrient is likely. Critical nutrient values may vary depending on the environment, growth conditions, and cultivar, which accounts for the range in values shown.

[c] Adequate nutrient levels represent either the minimum nutrient concentration for nonlimited growth derived from greenhouse studies or nutrient ranges measured in field-grown plants that were adequately supplied with the nutrient in question.

Table 10.4. Critical and toxic concentrations for micronutrients in cotton

Nutrient	Growth stage	Index tissue	Concentration (ppm)		Reference
			Critical[a]	Toxic[b]	
B	Seasonal	Leaf blade	10–28	>1,000	23, 37, 38, 54, 64, 65, 74
Mn	Seasonal	Leaf blade	10–15	2,000–3,700	24, 28, 67
Zn	31 DAP[c]	Leaf blade	11	200	68
Mo	65 DAP	Leaf blade	0.5	NE[d]	5
Fe, Cu	...	...	...	NE	...

[a] Critical concentration is the level below which a reduction in plant growth and expression of foliar deficiency symptoms occur.

[b] Toxic nutrient concentration thresholds represent accumulation levels that result in expression of foliar toxicity symptoms and a reduction in plant growth.

[c] Days after planting.

[d] Critical tissue deficiency or toxicity concentrations have not been established.

(*Alternaria macrospora* A. Zimmerm.) and Verticillium wilt (11,39). It has been stated (39) that expression of K deficiency symptoms is a prerequisite for an Alternaria epidemic and also that cultivar differences in susceptibility to Alternaria leaf spot are related to the severity of K deficiency symptoms when weather conditions are favorable for disease development.

Because the boll represents the primary sink for many essential nutrients, introduction of new technologies or changes in crop management that increase seed cotton yields often require changes in soil fertility management. Greater yields mean increased total plant nutrient requirements and greater nutrient removal with harvested seed cotton. In this regard, it is noteworthy that genetic improvement of modern cotton cultivars has been associated with increased harvest indexes and earlier fruiting (83,84). Development of narrow-row production systems with increased plant density is likely to further accelerate the trend towards cultivars with higher harvest indexes. Genotypes with shorter stature, shorter fruiting branches, and relatively less vegetative biomass (i.e., higher harvest index) appear to produce greater yields in narrow-row systems than the cultivars used presently (51,52). Such trends will require a renewed effort to update plant tissue nutrient guidelines and soil test recommendations.

Selected References

1. Adams, F., and Hathcock, P. J. 1984. Aluminum toxicity and calcium deficiency in acid subsoil horizons of two Coastal Plains soil series. Soil Sci. Soc. Am. J. 48:1305-1309.
2. Adams, F., and Lund, Z. F. 1966. Effect of chemical activity of soil solution aluminum on cotton root penetration of acid subsoils. Soil Sci. 101:193-198.
3. Adams, F., and Moore, B. L. 1983. Chemical factors affecting root growth in subsoil horizons of Coastal Plain soils. Soil Sci. Soc. Am. J. 47:99-102.
4. Adams, F., and Wear, J. I. 1957. Manganese toxicity and soil acidity in relation to crinkle leaf of cotton. Soil Sci. Soc. Am. Proc. 21:305-308.
5. Amin, J. V., and Joham, H. E. 1960. Growth of cotton as influenced by low substrate molybdenum. Soil Sci. 89:101-107.
6. Anderson, O. E., and Boswell, F. C. 1968. Boron and manganese effects on cotton yield, lint quality, and earliness of harvest. Agron. J. 60:488-493.
7. Bassett, D. M., Anderson, W. D., and Werkhoven, C. H. E. 1970. Dry matter production and nutrient uptake in irrigated cotton (*Gossypium hirsutum*). Agron. J. 63:399-403.
8. Bassett, D. M., and George, A. G. 1983. Plant analysis as a guide to cotton fertilization. Pages 30-31 in: Soil and Plant Tissue Testing in California. H. M. Reisenauer, ed. Univ. Calif. Div. Agric. Sci. Bull. 1879.
9. Bennett, O. L., Rouse, R. D., Ashley, D. A. and Doss, B. D. 1965. Yield, fiber quality, and potassium content of irrigated cotton plants as affected by rates of potassium. Agron. J. 57:296-299.
10. Braud, M. 1974. The control of mineral nutrition of cotton by foliar analysis. Coton Fibres Trop. 29:215-235.
11. Broome, J. C. 1990. Interactions of the pathogen, cotton roots, and soil environmental parameters on Verticillium wilt of cotton. M.Sc. thesis. University of California, Davis.
12. Brouder, S. M., and Cassman, K. G. 1990. Root development of two cotton cultivars in relation to potassium uptake and plant growth in a vermiculitic soil. Field Crops Res. 23:187-203.
13. Brown, J. C., and Jones, W. E. 1977. Fitting plants nutritionally to soils. II. Cotton. Agron. J. 69:405-409.
14. Burnbaum, E. H., Beasley, C. A., and Duggar, W. M. 1974. Boron deficiency in unfertilized cotton ovules grown in vitro. Plant Physiol. 54:931-935.
15. Cassman, K. G., Kerby, T. A., Roberts, B. A., Bryant, D. C., and Brouder, S. M. 1989. Differential response of two cotton cultivars to fertilizer and soil potassium. Agron. J. 81:870-876.
16. Cassman, K. G., Kerby, T. A., Roberts, B. A., Bryant, D. C., and Higashi, S. L. 1990. Effect of potassium nutrition on lint yield and fiber quality of Acala cotton. Crop Sci. 30:672-677.
17. Cope, J. T. 1981. Effects of 50 years of fertilization with phosphorus and potassium on soil test levels and yields at six locations. Soil Sci. Soc. Am. J. 45:342-347.
18. Cope, J. T. 1984. Relationships among rates of N, P, and K, soil test values, leaf analysis and yield of cotton at six locations. Commun. Soil Sci. Plant Anal. 15:253-276.
19. Cramer, G. R., Lynch, J., Lauchli, A., and Epstein, E. 1987. Influx of Na, K, and Ca into roots of salt-stressed cotton seedlings. Plant Physiol. 83:510-516.
20. Dhindsa, R. S., Beasley, C. A., and Ting, I. P. 1975. Osmoregulation in cotton fiber. Plant Physiol. 56:394-398.
21. Donald, L. 1964. Nutrient deficiencies in cotton. Pages 59-98 in: Hunger Signs in Crops. H. B. Sprague, ed. David McKay, New York.
22. Eaton, F. M. 1942. Toxicity and accumulation of chloride and sulfate salts in plants. J. Agric. Res. 64:357-399.
23. Eaton, F. M. 1944. Deficiency, toxicity, and accumulation of boron in plants. J. Agric. Res. 69:237-277.
24. Foy, C. D., Fleming, A. L., and Armiger, W. H. 1969. Differential tolerance of cotton varieties to excess manganese. Agron. J. 61:690-694.
25. Foy, C. D., Jones, J. E., and Webb, H. W. 1980. Adaptation of cotton genotypes to an acid, aluminum toxic soil. Agron. J. 72:833-839.
26. Gardner, B. R., and Tucker, T. C. 1967. Nitrogen effects on cotton: Vegetative and fruiting characteristics. Soil Sci. Soc. Am. Proc. 31:780-785.
27. Gerik, T. J., Morrison, J. E., and Chichester, F. W. 1987. Effects of controlled-traffic on soil physical properties and crop rooting. Agron. J. 79:434-438.
28. Gheesling, R. H., and Perkins, H. F. 1970. Critical levels of manganese and magnesium in cotton at different stages of growth. Agron. J. 62:29-32.
29. Grimes, D. W., Miller, R. J., and Wiley, P. L. 1975. Cotton and corn root development in two field soils of different strength characteristics. Agron. J. 67:519-523.
30. Gulick, S. H., Cassman, K. G., and Grattan, S. R. 1989. Exploitation of soil potassium in layered profiles by root systems of cotton and barley. Soil Sci. Soc. Am. J. 53:146-153.
31. Halevy, J. 1976. Growth rate and nutrient uptake of two cotton cultivars grown under irrigation. Agron. J. 68:701-705.
32. Halevy, J., and Bazelet, M. 1989. Fertilizing for high yield and quality, cotton (revised). IPI (Int. Potash Inst.) Bull. 2.
33. Halevy, J., Marani, A., and Markovitz, T. 1987. Growth and NPK uptake of high-yielding cotton grown at different nitrogen levels in a permanent-plot experiment. Plant Soil 103:39-44.
34. Harris, C. H., and Smith, C. W. 1980. Cotton production affected by row profile and N rates. Agron. J. 72:919-922.
35. Hearn, A. B. 1981. Cotton nutrition. Field Crops Abstr. 34:13-34.

36. Hearn, A. B. 1986. Effect of preceding crop on the nitrogen requirements of irrigated cotton (*Gossypium hirsutum* L.) on a vertisol. Field Crops Res. 13:159-175.

37. Heathcote, R. G., and Smithson, J. B. 1974. Boron deficiency in cotton in northern Nigeria. I. Factors influencing occurrence and methods of correction. Exp. Agric. 10:199-208.

38. Heathcote, R. G., and Smithson, J. B. 1974. Boron deficiency on cotton in northern Nigeria. II. The effect of variety. Exp. Agric. 10:209-218.

39. Hillocks, R. J., and Chinodya, R. 1989. The relationship between Alternaria leaf spot and potassium deficiency causing premature defoliation of cotton. Plant Pathol. 38:502-508.

40. Hinkle, D. A., and Brown, A. L. 1968. Secondary and micronutrients. Pages 281-320 in: Advances in Production and Utilization of Quality Cotton: Principles and Practices. F. C. Elliot, M. Hoover, and W. K. Porter, Jr., eds. Iowa State University Press, Ames.

41. Holley, K. T., and Dulin, T. G. 1939. Influence of boron on flower bud development in cotton. J. Agric. Res. 59:541-545.

42. Howard, D. D., and Adams, F. 1965. Calcium requirement for penetration of subsoils by primary cotton roots. Soil Sci. Soc. Am. Proc. 29:558-562.

43. Huber, D. M., and Arny, D. C. 1985. Interaction of potassium with plant disease. Pages 467-488 in: Potassium in Agriculture. R. D. Munson, ed. American Society of Agronomy, Madison, WI.

44. Hue, N. V., Craddock, G. R., and Adams, F. 1986. Effect of organic acids on aluminum toxicity in subsoils. Soil Sci. Soc. Am. J. 50:28-34.

45. Ishag, H. M., Ayoub, A. T., and Said, M. B. 1987. Cotton leaf reddening in the irrigated Gezira. Exp. Agric. 23:207-212.

46. Joham, H. E. 1951. The nutritional status of the cotton plant as indicated by tissue tests. Plant Physiol. 26:76-89.

47. Joham, H. E., and Amin, J. V. 1967. The influence of foliar and substrate application of manganese on cotton. Plant Soil 26:369-379.

48. Jones, U. S., and Bardsley, C. E. 1968. Phosphorus nutrition. Pages 212-253 in: Advances in Production and Utilization of Quality Cotton: Principles and Practices. F. C. Elliot, M. Hoover, and W. K. Porter, Jr., eds. Iowa State University Press, Ames.

49. Kapur, M. L., and Sekhon, G. S. 1985. Rooting pattern, nutrient uptake and yield of pearl millet (*Pennisetum typhoideum* Pers.) and cotton (*Gossypium herbaceum*) as affected by nutrient availability from the surface and subsurface soil layers. Field Crops Res. 10:77-86.

50. Kennedy, C. W., Ba, M. T., Caldwell, A. G., Hutchinson, R. L., and Jones, J. E. 1987. Differences in root and shoot growth and soil moisture extraction between cotton cultivars in an acid subsoil. Plant Soil 101:241-246.

51. Kerby, T. A., Cassman, K. G., and Keeley, M. 1990. Genotypes and plant densities for narrow row cotton systems. I. Height, nodes, earliness, and location of yield. Crop Sci. 30:644-649.

52. Kerby, T. A., Cassman, K. G., and Keeley, M. 1990. Genotypes and plant densities for narrow row cotton systems. II. Leaf area and dry matter partitioning. Crop Sci. 30:647-653.

53. Leffler, H. R., and Tubertini, B. S. 1976. Development of cotton fruit. II. Accumulation and distribution of mineral nutrients. Agron. J. 68:858-861.

54. Lombin, G. 1981. Approximating the potassium fertilization requirement of cotton on some representative semiarid tropical savannah soils of Nigeria. Can. J. Soil Sci. 61:507-516.

55. Lombin, G. 1983. Comparative tolerance and susceptibility of cotton and peanuts to high rates of Mn and B applications under greenhouse conditions. Soil Sci. Plant Nutr. 29:363-368.

56. Longenecker, D. E. 1974. The influence of high sodium in soils upon fruiting and shedding, boll characteristics, fiber properties, and yields of two cotton species. Soil Sci. 118:387-396.

57. Lunt, O. R., and Nelson, W. L. 1950. Studies on the value of sodium in the mineral nutrition of cotton. Soil Sci. Soc. Am. Proc. 15:195-200.

58. Maas, E. V., and Hoffman, G. J. 1977. Crop salt tolerance—Current assessment. J. Irrig. Drain. Div. Am. Soc. Civil Eng. 103 (IR2):115-134.

59. Mahler, R. J. 1989. Sulfur effects on cotton cultivars grown in a greenhouse. J. Plant Nutr. 12:187-206.

60. Malik, M. N. A., Evenson, J. P., and Edwards, D. G. 1978. The effect of level of nitrogen nutrition on earliness in upland cotton (*Gossypium hirsutum* L.). Aust. J. Agric. Res. 29:1213-1221.

61. Maples, R., and Keogh, J. L. 1963. Effects of boron deficiency on cotton. Ark. Farm Res. 12:5.

62. Maples, R., and Keogh, J. L. 1973. Phosphorus fertilization experiments with cotton on Delta soils of Arkansas. Ark. Agric. Exp. Stn. Bull. 781.

63. Marcus-Wyner, L., and Rains, D. W. 1982. Nutritional disorders of cotton plants. Commun. Soil Sci. Plant Anal. 13:685-736.

64. Miley, W. N., Hardy, G. W., Sturgis, M. B., and Sedberry, J. E. 1969. Influence of boron, nitrogen, and potassium on yield, nutrient uptake, and abnormalities of cotton. Agron. J. 61:9-13.

65. Murphy, B. C., and Lancaster, J. D. 1971. Response of cotton to boron. Agron. J. 63:539-540.

66. Oertli, J. J., and Roth, J. A. 1969. Boron nutrition of sugar beet, cotton, and soybean. Agron. J. 61:191-195.

67. Ohki, K. 1974. Manganese nutrition of cotton under two boron levels. II. Critical Mn level. Agron. J. 66:572-575.

68. Ohki, K. 1975. Lower and upper critical zinc levels in relation to cotton growth and development. Plant Physiol. 35:96-100.

69. Page, A. L., and Bingham, F. T. 1965. Potassium-magnesium interrelationships in cotton. Calif. Agric. 19(11):6-7.

70. Pluenneke, R. H., and Joham, H. E. 1972. The influence of low substrate sodium levels upon the free amino acid content of cotton leaves. Plant Physiol. 49:502-505.

71. Ramani, S., and Kannan, S. 1982. Inadaptive changes in pH with Zn stress tolerance in some cultivars of cotton and peanut. J. Plant Nutr. 5:207-217.

72. Rathert, G. 1982. Influence of extreme K:Na ratios and high substrate salinity on plant metabolism of crops differing in salt tolerance. J. Plant Nutr. 5:133-193.

73. Rios, M. A., and Pearson, R. W. 1964. The effect of some chemical environmental factors on cotton root behavior.

Soil Sci. Soc. Am. Proc. 28:232-235.

74. Rothwell, A., Bryden, J. W., Knight, H., and Coxe, B. J. 1967. Boron deficiency of cotton in Zambia. Cotton Grow. Rev. 52:293-308.

75. Silberbush, M., and Ben-Asher, J. 1987. The effect of salinity on parameters of potassium and nitrate uptake of cotton. Commun. Soil. Sci. Plant Anal. 18:65-81.

76. Skinner, J. J., and Allison, F. E. 1923. Influence of fertilizers containing borax on the growth and fruiting of cotton. J. Agric. Res. 23:433-441.

77. Stanford, G., and Jordan, H. V. 1966. Sulfur requirements of sugar, fiber, and oil crops. Soil Sci. 101:258-266.

78. Stromberg, L. K. 1960. Potassium fertilizer on cotton. Calif. Agric. 14(4):4-5.

79. Tabor, J. A., Pennington, D. A., and Warrick, A. W. 1984. Sampling variability of petiole nitrate in irrigated cotton. Commun. Soil Sci. Plant Anal. 15:573-585.

80. Thompson, A. C., Lane, H. C., Jones, J. W., and Hesketh, J. D. 1976. Nitrogen concentrations of cotton leaves, buds, and bolls in relation to age and nitrogen fertilization. Agron. J. 68:617-621.

81. Tupper, G. R., and Ebelhar, M. W. 1990. Fertilizer placement: Past, present, and future? In: Proc. Beltwide Cotton Production Res. Conf. J. M. Brown, ed. National Cotton Council, Memphis, TN.

82. Vretta-Kouskoleka, H., and Kallinis, T. L. 1968. Iron deficiency in cotton in relation to growth and nutrient balance. Soil Sci. Soc. Am. Proc. 32:253-257.

83. Wells, R., and Meredith, W. R., Jr. 1984. Comparative growth of obsolete and modern cotton cultivars. I. Vegetative dry matter partitioning. Crop Sci. 24:858-862.

84. Wells, R., and Meredith, W. R., Jr. 1984. Comparative growth of obsolete and modern cotton cultivars. II. Reproductive dry matter partitioning. Crop Sci. 24:863-868.

85. Weir, B. L., Kerby, T. A., Roberts, B. A., Mikkelsen, D. S., and Garber, R. H. 1986. Potassium deficiency syndrome of cotton. Calif. Agric. 40(5-6):13-14.

86. Wiles, A. B. 1959. Calcium deficiency in cotton seedlings. Plant Dis. Rep. 43:365-367.

Part III

Vegetable Crops

Chapter 11

Salvadore J. Locascio
Vegetable Crops Department, Institute of Food and Agricultural Sciences
University of Florida, Gainesville

Cucurbits: Cucumber, Muskmelon, and Watermelon

Cucumber (*Cucumis sativus* L.), muskmelon (*Cucumis melo* var. *reticulatus* L.), and watermelon (*Citrullus lanatus* (Thumb.) Matsum & Nakai) are the most widely grown vegetables of the Cucurbitaceae family. These warm-season crops are intolerant of freezing temperatures and grow best at relatively high temperature. They are vining crops with similar cultural requirements and are grown extensively in home gardens and for commercial production for their fruit. Cucumbers are commonly used in salads and for making pickles, whereas muskmelon and watermelon are grown for their sweet fruit.

Optimum seed germination occurs at a soil temperature of 77–95° F (25–35° C) with greatly inhibited germination at temperatures below 60° C (15° F). Seed germination is very rapid (4–5 days) at high temperatures but may take 15–20 days at low temperatures. The number of days from seeding to maturity depends on the cultivar grown (fruit size) and soil and air temperatures. Cucumber commonly requires 40–60 days, and muskmelon (commonly called cantaloupe) requires 70–85 days. Large-fruited watermelon cultivars require 100–120 days, but small-fruited cultivars mature in 75–100 days. Transplants and polyethylene mulch are used to enhance earliness in watermelon and muskmelon.

These crops grow most abundantly with adequate soil moisture and are generally irrigated for maximum production. Watermelons have an extensive root system and are more drought tolerant than cucumber or muskmelon. Cucurbits are very sensitive to water-logged soil and on heavy, poorly drained soils are grown on beds to enhance drainage. These crops produce male and female flowers and require bees for pollination and fruit set.

The main cucurbit commercial production centers are located in the southern United States and California. Slicer cucumber production is highest in Florida, followed by Texas and California. Processing cucumber acreage is highest in Michigan, North Carolina, and Ohio. Muskmelon commercial production is concentrated in arid or semiarid regions. California produces about 70% of muskmelon grown in the United States, and Texas and Arizona are also major production areas. Florida leads the United States in watermelon production with about 30% of the crop annual acreages on about 50,000 acres (20,235 ha). Texas and California each produce about one-half the amount of Florida production. The trend in production is for reduced acreages and increased management intensity resulting in an increase in total production from the acreage.

Deficiency Symptoms

Nitrogen (N)

Cucurbits grown with an inadequate N supply (less than 2–2.5% dry weight) exhibit easily recognized deficiency symptoms. The vegetative growth rate is reduced, and a yellowing of the leaves due to a loss in chlorophyll will be observed. The loss of the green color occurs first on the more mature leaves and last on the younger leaves because of the degradation of N-containing compounds in the older leaves and movement of N to the young tissue. The size of the plant canopy is reduced. The smaller plant results in a reduced carbohydrate assimilation. Fruit set is reduced, and developed fruit are small and subject to sunburn (Plate 161). Muskmelon grown with N deficiency produce fruit with low netting and poor general appearance, and plants have a high incidence of cull fruit. The reduction in yield and quality are directly related to the severity of the N deficiency.

Phosphorus (P)

Foliar symptoms of P deficiency are often difficult to recognize. Cucurbits have an overall stunted appearance and a slower growth rate. Where N is sufficient, the plant will be a dark green. Under cool growing conditions, leaves may be slightly purple. Visual symptoms are not always present even though the cucurbit would respond to additional P. A concentration of P below 0.2% in the recently matured leaves on mature plants is associated with P deficiency.

Potassium (K)

The first symptom of K deficiency is slow growth. If the deficiency continues, the tips of the older leaves develop chlorosis and later become necrotic (Plate 162). Fruit development is irregular, with reduced growth at the stem end and normal to enlarged growth at the blossom end. Because N accumulates in K-deficient tissue, the incidence of disease may increase because of an increase in N available for microorganism growth. Thus, adequate K results in maximum growth and a lesser incidence of some diseases.

Calcium (Ca)

With a severe Ca deficiency, new root growth is impaired, and this further reduces nutrient uptake. Foliar symptoms occur on new leaves, since Ca in the plant is not mobile. Young leaves are reduced in size, distorted, and spotted or necrotic at the leaf tips. Under conditions of severe Ca deficiency, terminal buds die. Such conditions in the field are uncommon, and most visible Ca deficiency symptoms occur on or affect the cucurbit fruit. In fact, leaf Ca concentrations and leaf appearance are a poor indication of Ca deficiency in fruit. In Ca-deficient fruit of watermelon and cucumber, it has been shown that the blossom end of the fruit wall becomes thinner, and water-soaked, brown necrotic spots develop. Because of the low transpiration rate of the fruit, the Ca concentration is lower than in the leaves. It has also been shown in cucumber fruit that a Ca concentration gradient exists; Ca is higher at the stem than at the blossom end.

Calcium deficiency problems can occur in cucurbits grown on a soil with relatively high concentrations of Ca. Uptake of Ca is influenced by a number of factors, the most important being soil water availability and the concentration and ratio of competing cations. In a dry soil, the growth of new roots required for Ca uptake is retarded, resulting in reduced Ca uptake. Also, as soils become dry, the concentration of soluble salts may increase, and high concentrations of soluble salts reduce Ca uptake. Potassium and NH_4^+, two cations that are often added in large quantities in fertilizer, compete with Ca for plant uptake. Maintaining the soil at an adequate water concentration and avoiding application of excessive K and ammonium-N in the fertilizer increases plant Ca uptake. Fruit that develops under conditions of Ca-stress may appear normal. However, after harvest the incidence of decay may be higher than in normal fruit.

Magnesium (Mg)

Symptoms of Mg deficiency occur first on older leaves, generally late in the growing season, and are associated with Mg concentrations below 0.2%. Chlorosis begins at the leaf tips and progresses to the interveinal area. The veins remain green, but the interveinal areas become yellow (Plate 163) before dying. Under severe stress conditions, fruit development and production are reduced. However, Mg deficiency that develops late in the growing season generally has little effect on fruit yield since older leaves normally are less metabolically active. Deficiencies occur most often on coarse-textured, acidic soils in humid regions. High concentrations of K^+, Ca^{2+}, and NH_4^+ compete with Mg^{2+} for uptake and may induce Mg deficiency.

Sulfur (S)

Symptoms of S deficiency are first observed on the young leaf tissue and eventually affect the entire plant. Deficiency is expressed by a reduction in green color. Plants are stunted, and the veinal area becomes slightly lighter than the interveinal area. The deficiency resembles that of N deficiencies except that S deficiency symptoms are observed first on the new growth. The occurrence of S deficiency under field conditions is becoming more common because low-S, high-analysis fertilizers are being used, and there is less S in the air due to stricter emissions laws.

Iron (Fe)

Iron is not readily mobile in the plant, and deficiency symptoms occur first on recently developed leaf tissue. Younger leaves become chlorotic. Larger veins remain green in the early stages of the deficiency, but later the entire leaf becomes uniformly chlorotic. In later stages, necrotic spots develop in the leaves.

Zinc (Zn)

Zinc deficiency occurs on new growth and is expressed as reduced leaf size, interveinal chlorosis that later becomes necrotic, and shortened internodes. Plant growth is reduced in proportion to the severity of the Zn deficiency. Deficiency is indicated by mature leaf Zn concentrations below 20–25 ppm. Zinc deficiencies are more severe when cucurbits are grown under cool, wet, and low-light-intensity situations. Also, excesses of other elements, particularly high soil-P concentrations, Fe, and Cu may induce Zn deficiency when Zn concentrations are low.

Manganese (Mn)

Because Mn is relatively immobile in the plant, symptoms of Mn deficiency occur on the new growth. The interveinal area of younger leaves becomes chlorotic while veins remain green. As the deficiency continues, older leaves become chlorotic. Manganese concentrations below 25–50 ppm indicate a deficiency of Mn. Manganese deficiencies can be induced by overliming.

Copper (Cu)

Under extreme Cu deficiency conditions, seeds germinate but growth rates are retarded, leaves are crinkled, internodes are shortened, and leaves become chlorotic and necrotic (Plate 164). In watermelons, vine growth may be limited to 1 ft (30 cm) compared with normal vine growth of 6–8 ft (1.8–2.4 m). On less deficient soils, early plant growth may appear normal. Later, when the plant's Cu needs increase, younger leaves may be distorted and the apical may die back. Tissue Cu concentrations of less than 2–3 ppm are associated with deficiency.

Boron (B)

Boron deficiency symptoms occur on young foliage and are characterized by shortened internodes and distorted leaves that become chlorotic, and then necrotic, with death of terminal buds. Fruit may be affected by cracks, necrotic spots, and internal breakdown.

Molybdenum (Mo)

Deficiencies of Mo are observed most often on strongly acidic soils with pH 5.0 or lower. Deficiencies of Mo interfere with plant N metabolism, specifically nitrate reduction to ammonium. Deficiency symptoms of Mo are similar to those of N deficiency and may be reduced by adding NH_4^+-N. In the early stages, older leaves become pale green with some interveinal mottling and later become necrotic. Fruit set is reduced and younger leaves become cupped, twisted, and distorted. Tissue Mo concentrations of 0.1–0.5 ppm are associated with deficiencies.

Chlorine (Cl)

Although Cl is an essential micronutrient, it has never been reported to be deficient in cucurbits under field conditions. Plants usually contain 50–500 ppm of Cl, with 2–20 ppm being associated with deficiencies. Most soils have abundant concentrations of Cl. Chlorine is added in the fertilizer, as in potassium chloride.

Toxicity Symptoms

Nitrogen

Excessive N is very detrimental to the production of cucurbit crops. With the excessive vegetative growth associated with excessive N, flowering is delayed and fruit set and yield are reduced. Since the visual response and increase in yield potential to added N can be quite dramatic, and the cost of fertilizer is relatively low, growers tend to apply N at or above recommended rates.

Potassium

Mature cucurbit plants have the capacity to absorb large quantities of K without deleterious effects. However, excessive K applied too close to the seed or transplant will result in soluble salt injury. Younger leaves wilt, and the leaf tip becomes necrotic. When fertilization is excessive, injury is greater with K from potassium chloride than K from potassium sulfate or potassium nitrate applied at the same rates. If injury

is observed at an early stage of growth, overhead irrigation can be used to leach excessive salts and minimize the injury.

Calcium

Toxicity of Ca does not commonly occur. Plants grow normally in calcareous soils such as marls, which are predominantly calcium carbonate. Toxic effects from Ca are mostly related to the accompanying anion. Excessive calcium chloride or calcium sulfate can result in soluble salt injury to cucurbits. Excessive soil Ca can increase soil pH to a level that reduces the availability of P and some micronutrients and result in reduced plant growth.

Magnesium

High rates of Mg do not generally directly cause plant growth problems. However, high Mg may reduce uptake of other elements, such as K, Ca, or Mn, and induce their deficiency. Mn deficiency can be induced by over-liming with dolomite or calcite.

Sulfur

Field-grown cucurbits are seldom affected by excess S in the soil. A symptom of excess soil S would be expressed as soluble salt injury. Excessive SO_2 gas in the atmosphere (0.5–0.7 mg SO_2-S/m^3) would result in leaf injury. Younger leaves are more tolerant to S dust than older leaves. Cultivar differences in S tolerance have been shown for muskmelon.

Zinc

Zinc toxicity in cucurbits is not widespread. Zinc concentrations in mature leaves of 400 ppm dry weight or greater are toxic. Plant root and top growth are reduced by toxicity. Zinc toxicity may result from high fertilization with Zn for a crop requiring high Zn, such as sweet corn, that is followed by a cucurbit crop, which has a lower Zn requirement.

Manganese

When growing cucurbits on high-Mn soils, particularly if the soil is low in Mg and Ca has a pH below 5, Mn toxicity can be severe. Symptoms of Mn toxicity occur with concentrations of 600–900 ppm. Seedlings germinate, but growth is slow, necrotic spots develop on the leaves, and the seedling may die. With less severe toxicity, plants grow slowly, and small distinct blackish brown spots develop on the underside of mature leaves. Preplant liming of these soils corrects this problem.

Copper

Copper is highly immobile in the soil, and applied Cu is either utilized by the plant or retained in the upper soil horizon. Excessive amounts of total Cu in the soil (200 ppm on a coarse-textured soil) may result in Cu toxicity. Copper is strongly bound to the root surface and reduces Fe uptake. Liming the soil reduces Cu toxicity. Since many fungicides commonly used for cucurbits contain Cu, soil-applied Cu may not be needed. In areas that are continually used for vegetables, care must be taken to avoid excessive Cu accumulation.

Boron

Application of excessive B results in toxicity and is characterized by yellowing of the leaf tips that progresses to the entire leaf. The chlorotic color progresses to necrosis, and the leaf has a burned appearance. Leaf B concentrations of over 200 ppm are associated with toxicity.

Molybdenum

Cucurbits are highly tolerant of excessive Mo. Molybdenum concentrations of 1,000 ppm or higher occur without visual toxic effects.

Chlorine

Chlorine toxicity is a more serious problem than deficiency in cucurbits and results in soluble salt injury. Symptoms include burning of leaf tips or margins, bronzing, and premature yellowing of leaves. Chlorine is highly mobile in the soil, and excessive concentrations can be leached from the soil by overhead irrigation.

Management Considerations

Cucurbits are grown successfully on a wide range of soil types. For early-season production, coarse-textured, deep, well-drained soils that warm up rapidly are preferred. Heavier soils can be used successfully with proper management, but early growth rates are generally slower than on lighter soils.

Cucurbits can be grown on soils with a wide range of acidity from pH 6.0 to 8.0. Watermelons are somewhat more tolerant of acidity than cucumbers and muskmelon and can be grown on soils within a range of pH 5.5–8.0. It is generally recommended to lime soil with pH values below 5.5 for watermelon and below 6.0 for cucumber and muskmelon. Many sandy soils typically used are very acidic when brought into cultivation. The pH of recently cleared, poorly drained sandy soils may be as low as 3.5. Better drained soils tend to be less acid. A soil test is essential on all land where the acidity and nutrient status are not known. A soil test is also necessary to determine the Ca and Mg status to determine the type of lime to apply. Cucurbits seeded on very acid soil will germinate, but generally the seedlings will die before the second or third true leaves develop. Such acid soils respond very favorably to lime.

Lime must be thoroughly incorporated to neutralize the soil acidity in the root zone. Generally, it is desirable to apply lime 1–2 months before seeding. However, studies have shown that the application of lime on coarse-textured soils a day before seeding provided results as beneficial as where lime was applied 1 month earlier. Lime should be applied broadcast and worked into the soil to a depth of 6–8 in. (15–20 cm) for a maximum growth and yield response. Application of lime only in the row or at very shallow depths will generally reduce the response to lime.

Crop rotation is desirable to reduce problems from soilborne diseases (mainly Fusarium wilt) and nematodes. For watermelon, the use of newly cleared land or land that has not been in watermelon production for 6–8 years is ideal. Rotations of such long duration for Fusarium wilt are not required for cucumber and muskmelon or where Fusarium-wilt-resistant watermelon cultivars are grown. When newly cleared land is not available in adequate amounts, land that previously has been in grass pastures is preferred over that previously used for continuous row crop production. Land in grass pasture is generally relatively free from weeds, and the extensive grass root system provides organic matter to improve soil tilth.

Nutrient Requirements

Soil tests provide a reliable indication of crop needs for P, K, secondary nutrients, and some micronutrients. Requirements for N and certain micronutrients currently are not as predictable by soil test. Since cucurbits are generally grown on coarse-textured and often recently cleared soils or soils that have not been in continuous cultivation, application of fertilizer that contains macronutrients and micronutrients is generally required for maximum production.

In the past, growers used organics in the fertilizer because yields for watermelon and other cucurbits were reported to improve with organic sources of N. Maximum watermelon production was limited until it was found that this response was to micronutrients supplied by the organic N source. Studies in Florida found Cu to be largely responsible for the yield response associated with natural organic N sources.

Fertilizer placement has a drastic effect on fruit yield of cucurbits. With the wide row spacings commonly used for watermelon and muskmelon, application of fertilizer in bands 2 in. (5 cm) to the side and 2 in. (5 cm) below the level of the seed or in a broad band in the center of the bed 2–4 in. (5–10 cm) under the seed often reduces yield from soluble salt injury. Soluble salt injury is more likely to occur with high fertilizer rates when irrigation is not used and with wide row spacings. Fertilization in a modified broadcast application and thorough incorporation of the fertilizer in the bed area 6–8 in. (15–20 cm) deep provides less soluble salt injury, reduces nutrient leaching, and results in much higher fruit yields than banded application. Where soluble N sources are used, one-half of the fertilizer should be applied in the bed before planting. The remaining N, P, and K should be applied in 12-in. (30-cm) wide bands on the bed shoulder and cultivated into the soil at lay-by. In years when rains cause excessive leaching, additional N and K should be applied and cultivated into the soil on the bed shoulders. Where black polyethylene mulch is used, all of the basic fertilizer can be applied broadcast in the bed area before application of the mulch. Where drip irrigation is used with polyethylene mulch, at least 60% of the N and K should be applied throughout the growing season through the drip system.

Cucurbits grow moderately well over a wide range of nutrient concentrations (Table 11.1). Maximum growth, however, is dependent on a number of factors including the availability of adequate nutrient concentrations.

Nitrogen

Nitrogen is the most abundant element in the plant after carbon, hydrogen, and oxygen. In general, this element is the most limiting factor for optimum growth in soils (other than organic soils) and generally gives the greatest response when applied. Nitrogen concentrations are greater in leaves of young plants than in leaves of older plants. Total N in the roots is less than that in the aboveground tissue. The N concentrations in normal leaves vary from 2.5 to 5% (Table 11.1). Plant vegetative growth is closely related to the N supply.

Cucurbits grown on mineral soils almost always respond to N fertilization with increased vegetative growth and intensity of green in the foliage. Nitrogen fertilization increases cell division and cell size, and this results in larger leaves and larger plants that produce more and larger fruit. The maturity of watermelon and muskmelon fruit are delayed slightly, because it takes longer for larger fruit to reach maturity.

Nitrogen requirements vary from 60 to 165 lb/acre (65 to 185 kg/ha) depending on the soil type, available

Table 11.1. Adequate nutrient concentrations for cucumber, muskmelon, and watermelon[a]

Element	Range
N, %	2.5–5.0
P, %	0.2–0.6
K, %	2.0–6.0
Ca, %	1.0–2.0
Mg, %	0.3–0.6
S, %	0.3–0.5
Fe, ppm	30–150
Zn, ppm	50–150
Mn, ppm	100–200
Cu, ppm	5–10
B, ppm	80–100
Mo, ppm	0.5–1.0
Cl, ppm	50–100

[a] For fully matured, whole leaf that is still metabolically active. Values are on a dry weight basis.

water, and length of growing season (Table 11.2). The lower rate is adequate on more fertile soils, for crops grown during a cooler period, and on soils where supplemental irrigation is not used.

Nitrogen can be supplied from ammonium nitrate, urea, ammonium sulfate, calcium nitrate, and potassium nitrate. Part of the N can be supplied from slow-release sources such as sulfur-coated urea and isobutylidene-diurea. Animal manure is an excellent source of N. In addition to supplying nutrients, organic matter increases the water-holding capacity of the soil and soil tilth. If available at a low cost, manure can be used to supply some of the plant nutrient requirements. Rates vary with the source. Cucurbits are very sensitive to ammonia. Excessive amounts of anhydrous ammonia and excess rates of manure are dangerous and may result in ammonium toxicity. If soil pH is too low and the soil temperature is low, nitrification of ammonium N sources may be slow. Under these conditions, 25–30% of the N should be in the nitrate form.

Phosphorus

Phosphorus is likely to be the second most limiting element after N for maximum cucurbit growth. The concentration of P in normal leaf tissue varies from 0.2 to 0.6% of dry weight. The P concentrations are higher in actively growing leaf tissue than in mature leaves, and P is also concentrated in seed.

Phosphorus stimulates root growth, thus cucurbits benefit from P applied early in the growing season, whereas sidedress P is less effective. In soils with high P fixation, banding the P near the seed is effective. On coarse-textured soils, P can be efficiently broadcast in the row before seeding.

Soil test results should be used to determine the P requirements of cucurbits. On a typical coarse-textured soil in Florida that tests low in P, yield response to application of P can be expected up to 160 lb/acre (180 kg/ha) as P_2O_5 (Table 11.2). On soils moderate in P, the required P rate is 100 lb/acre (110 kg/ha) as P_2O_5, and no P should be applied to soils testing very high in P.

Phosphorus can be supplied from ordinary superphosphate or concentrated superphosphate. In studies with banded P from diammonium phosphate, yields of watermelon and cucumber were lower than with the first two sources of P. Diammonium phosphate increases soil pH, and its application in bands may reduce micronutrient uptake, resulting in reduced yields. This effect is minimized by broadcast application and by using diammonium phosphate as only part of the P.

Potassium

Potassium concentrations in cucurbits range from 2 to 6% and are higher in young plants than mature plants. Potassium is highly mobile in the plant and is translocated from the leaf tissue into fruit. Under

Table 11.2. Fertilizer rates recommended for cucurbits on soils testing low for P and K[a]

Crop	N lb/acre	N kg/ha	P₂O₅ lb/acre	P₂O₅ kg/ha	K₂O lb/acre	K₂O kg/ha
N			P_2O_5		K_2O	
	lb/acre	kg/ha	lb/acre	kg/ha	lb/acre	kg/ha
Florida[b,c]						
Cucumber	90	101	120	134	120	134
Muskmelon	120	134	160	179	120	134
Watermelon	120	134	160	179	120	134
Indiana[b]						
Cucumber	90	101	60	67	120	134
Muskmelon	90	101	60	67	120	134
Watermelon	60	67	60	67	120	134
Texas[d,e]						
Cucumber	120	134	100	112	0–60	0–67
Muskmelon	120	134	100	112	0–60	0–67
Watermelon[f]	60	67	80	90	0–40	
New York[b]						
Cucumber	80–165	90–185	50–150	56–168	50–150	56–168
Muskmelon	80–165	90–185	50–150	56–168	50–150	56–168
Watermelon[f]	80–165	90–185	50–150	56–168	50–150	56–168
California[g]						
Cucumber	121	136	60	67	24	27
Muskmelon	121	136	60	67	54	60

[a] Adapted from the *Handbook for Vegetable Growers* (8).
[b] Apply all P and one-half N and K broadcast in the bed at planting, with remaining N and K in one or two supplemental applications.
[c] An additional 30:0:30 can be applied if excessive rainfall occurs.
[d] Apply all P in the bed at planting with one-half N and K applied when vines start to run and one-half following full bloom.
[e] Most soils test high in K, and no K is required.
[f] Not irrigated.
[g] Mean fertilizer quantities applied by growers. On soil testing low in P and K, requirements are higher.

conditions of K stress, K moves from the older leaves to the younger tissue; thus, visual symptoms of K deficiency first develop on the older leaves. Potassium can be supplied from potassium chloride, potassium sulfate, potassium nitrate, or potassium-magnesium sulfate.

Although most soils contain some K, rapidly growing cucurbits generally require additions of K for maximum growth. Soil test results are needed to determine the K requirements on a specific soil. Highly leached coarse-textured soils and new organic soils are deficient in K. However, it is not advisable to grow these cucurbits on organic soils, because mineralization of the organic matter may result in the availability of excessive N. In arid regions, K leaching is less severe; thus, K may be less limiting where rainfall is low. In Florida, K requirements vary from 0 to 120 lb/acre (0 to 135 kg/ha) K_2O, and the rate should be determined from soil test results. The K requirements of cucurbits grown on coarse-textured soils testing very low in K in Florida are shown in Table 11.2.

Calcium

Plant calcium requirements are about 1–2% in mature leaf tissue and 0.2–0.3% in cucurbit fruit. Tissues that have a higher transpiration rate such as leaves are higher in Ca concentration than tissues such as fruit and stems that have a lower transpiration rate. In a fertile soil, Ca is the predominant cation and should constitute at least 60–85% of the cation exchange capacity (CEC). In soils where Ca makes up less than 25–30% of the total CEC, Ca-related nutritional problems are likely to occur on cucurbits.

Soils that are low in Ca are also generally acidic. The Ca and pH status of a soil should be determined by soil test. Where Ca is needed and the pH is low, lime is the least expensive source of Ca. Lime sources include calcium carbonate, calcium hydroxide, and calcium oxide. Where soil Mg is also low, a dolomitic limestone ($Ca \cdot MgCO_3$) should be used. Soils should be limed if the pH is below 5.5 for watermelon and below 6.0 for cucumber and muskmelon. If soil Ca contents are low (Ca less than 40–50% of the CEC) and the pH is in the proper range, gypsum can be applied to supply Ca.

Magnesium

Magnesium concentration in healthy cucurbit leaf tissue ranges from 0.3 to 0.6%. Magnesium is highly mobile in the plant, and a continuous supply is necessary to avoid Mg stress and development of deficiency symptoms.

Preplant soil test results should be used to determine

the soil Mg status. Soils that are below 25–50 ppm exchangeable Mg should be fertilized with Mg. If the pH is also low, Mg can be supplied from dolomite. If the soil pH is in the proper range, other Mg sources include magnesium oxide and magnesium sulfate.

Foliar sprays of magnesium sulfate can be used to alleviate Mg deficiency; however, preplant soil applications of Mg where needed are more effective, since several applications would be needed if applied by foliar sprays.

Sulfur

Plants can absorb S by roots and leaf stomata. Uptake is mainly as SO_4^{2-} by roots and as SO_2 gas by the leaves. Sulfur does not translocate from older leaves to newer leaves, so a continuous supply is necessary for maximum growth. Normal leaf tissue of cucurbits contains 0.3–0.5% S on a dry weight basis.

Most of the S in soils exists in the organic matter, and this is converted to inorganic SO_4^{2-} by microbial breakdown. The SO_4^{2-} is soluble and leaches readily under high-rainfall conditions but accumulates in the upper soil layers in arid regions. The major sources of atmospheric SO_2 are from industry, auto emissions, and the ocean. In addition, S is commonly supplied in many insecticides, fungicides, fertilizers, and some irrigation water sources. Because of the numerous sources of S, the use of S fertilizers has been minimal in the past. With increased control of SO_2 emission, higher analysis fertilizers with less S, and the advent of organic pesticides, S deficiencies are occurring more often, and S fertilizer use is increasing. Where S deficiency has been diagnosed or is expected, S-containing fertilizers should be used. Where needed, S should be added at about 20–40 lb/acre (22–45 kg/ha). Where the pH is high, elemental S can be used to decrease soil pH and provide the plant nutrient.

Iron

Iron in normal cucurbit leaf tissue ranges from 30 to 150 ppm on a dry weight basis and may be as high as 500 ppm without toxicity. Most soils contain large quantities of Fe, but plant availability is strongly pH dependent, with minimum availability at pH 7.8–8.5. Since cucurbits are generally grown on soils of pH 6.5 or lower, Fe deficiencies do not occur often. Soil testing should be used to accurately determine lime requirements to avoid overliming and lime-induced Fe deficiency. Many cations including Cu, Zn, Mn, K, Ca, and Mg compete with Fe for absorption. Where the soil pH is 7.0 or higher, it can be reduced with elemental S, or acid-forming fertilizers such as ammonia, ammonium sulfate, or ammonium thiosulfate can be applied. Soil-applied Fe can be supplied as ferrous sulfate at 4 lb/acre of Fe (5 kg/ha) or Fe-chelates sprayed on foliage at 1–2 lb/acre (1.1–2.2 kg/ha).

Zinc

The concentration of Zn in cucurbit leaf tissue is about 50–150 ppm. In native soils, Zn concentrations are usually higher at the soil surface because they are released from accumulated organic matter more than from subsurface soil. Zinc is more available in an acid soil, with reduced availability at pH 6–7. Even at lower pH values, available Zn can be immobilized in the soil and become unavailable. Zinc deficiency is very common on coarse-textured and fine, silty-loam soils in highly leached humid regions; in washed river bottom soils; and in calcareous soils. Many soils commonly used to grow cucurbits in Florida, California, Louisiana, Texas, and Oklahoma are deficient in Zn.

Soil test results can determine the Zn status of a soil. Where Zn deficiency is expected, Zn can be applied at 3–10 lb/acre (3.3–11 kg/ha), depending on the soil type, from zinc sulfate, zinc oxide, or zinc chelate. When growing cucurbits on soils of pH 7 or higher, several applications of a foliar Zn spray each of 1–2 lb/acre of Zn (1.1–2.2 kg/ha) are effective. Many fungicides commonly used on cucurbits contain Zn, and these applications also effectively supply Zn as a nutrient.

Manganese

The Mn content of normal cucurbit leaf tissue is about 100–200 ppm. The range of Mn in soils varies from as low as 5 ppm to as high as 5,000 ppm. Availability is highly pH dependent, with low availability at pH 6.5 and higher. Therefore, Mn deficiencies occur commonly on coarse-textured soils with low Mn content and on alkaline soils.

Manganous sulfate is the most widely used Mn source, and it can be applied to the soil or as a foliar spray. The rate of Mn application on a deficient coarse-textured soil is 5–10 lb/acre of Mn (6–11 kg/ha), whereas a higher rate may be required on calcareous soils.

Copper

Cucurbits are highly responsive to Cu when they are grown on soils where amounts of available Cu are low. Copper requirements are about 5–10 ppm on a dry weight basis. The total Cu content of most agricultural soils ranges from 20 to 250 ppm. Soil deficiencies have been reported in numerous states. Copper is very strongly bound by soil organic matter. Organic soils are generally Cu deficient when brought into cultivation. Also, some highly weathered and coarse-textured soils, such as the soils of the Atlantic coastal plains, are much lower in Cu, ranging from 1 to 30 ppm (extractable). Many of these soils are very acid, as low as pH 3.5, when brought into cultivation. Liming that is required for good crop production reduces Cu availability. Copper availability is drastically reduced at a pH above 7 and is most available below pH 6. When Cu deficiencies are expected, Cu can be supplied at 3–4 lb/acre (3.4–5 kg/ha) in the fertilizer from copper sulfate or copper oxide. Broadcast application of Cu with a fertilizer has resulted in higher cucumber and watermelon production than band placement. Copper can also be supplied as a foliar spray at 0.25 lb/acre (280 g/ha), but numerous applications are generally required to correct a deficiency, because a continued supply is needed throughout the season.

Boron

Boron concentration in normal cucurbit tissue is about 80–100 ppm of B on a dry weight basis. Cucurbits have a moderate B requirement. Boron deficiencies are widespread, and applications are most likely required on soils that are coarse-textured, highly leached, and alkaline. The concentration of B in most soils ranges from 20 to 200 ppm but hot-water-soluble (available) B ranges from 0.4 to 5 ppm. Less than 1 ppm of hot-water-soluble B is considered deficient, whereas above 5 ppm is toxic. Boron is most available at low pH (4.7) and decreases to pH 6.7. Boron is commonly applied at 0.5–3 lb/acre of B (0.55–3.3 kg/ha) in the form of sodium borate. Soluble boron is leached and does not accumulate under humid or irrigated conditions.

Molybdenum

Of all the essential elements required by cucurbits, Mo is needed in the lowest amounts, 0.5–1.0 ppm (Table 11.1). The content of total Mo in most mineral soils is extremely low at 0.6–3.5 ppm, but this is apparently adequate if the soil is limed to pH 5.5 or above. A Mo deficiency can be corrected by liming acidic soil or adding 5 oz/acre of Mo (350 g/ha) as sodium molybdate or ammonium molybdate.

Selected References

1. Elamin, O. M., and Wilcox, G. E. 1986. Effect of soil acidity and magnesium on muskmelon leaf composition and fruit yield. J. Am. Soc. Hortic. Sci. 111:682-685.

2. Elamin, O. M., and Wilcox, G. E. 1989. Effect of magnesium and manganese nutrition on watermelon growth and manganese toxicity. J. Am. Soc. Hortic. Sci. 114:588-593.

3. Frost, D. J., and Kretchman, D. W. 1989. Calcium deficiency reduces cucumber fruit and seed quality. J. Am. Soc. Hortic. Sci. 114:552-556.

4. Johnson, J., Jr., and Mayberry, K. S. 1980. The effect of dusting sulfur on muskmelons. HortScience 15:652-654.

5. Locascio, S. J., and Fiskell, J. G. A. 1966. Copper requirement of watermelon. Proc. Am. Soc. Hortic. Sci. 88:568-575.

6. Locascio, S. J., Fiskell, J. G. A., and Everett, P H. 1970. Advances in watermelon fertility. Proc. Trop. Reg. Am. Soc. Hortic. Sci. 14:223-231.

7. Locascio, S. J., Fiskell, J. G. A., and Martin, F. G. 1972. Influence of fertilizer placement and micronutrient rate on watermelon composition and yield. J. Am. Soc. Hortic. Sci. 97:1919-123.

8. Lorenz, O. A., and Maynard, D. N. 1988. Handbook for Vegetable Growers. 3rd ed. John Wiley & Sons, New York.

9. Navarro, A. A., and Locascio, S. J. 1979. Copper nutrition of cucumber (*Cucumis sativus* L.) as influenced by fertilizer placement, phosphorus rate, and phosphorus source. Soil Crop Sci. Soc. Fla. Proc. 39:16-19.

10. Shear, C. B. 1975. Calcium related disorders of fruits and vegetables. HortScience 10:361-365.

Chapter 12

David A. Bender
Texas Agricultural Experiment Station
Lubbock, Texas

Onions

The bulb onion (*Allium cepa* L.) and its close relatives, leeks (*A. ampeloprasum* L.), Japanese bunching onions (*A. fistulosum* L.), garlic (*A. sativum* L.), and a few others are unique among vegetables. Few other vegetables are monocots, and only members of the genus *Allium* are grown for their edible bulbs or leaf bases. Onions have shallow root systems consisting of un-branched secondary roots that are frequently lost and replaced during the growing season. Thus, the relatively high water and nutrient requirements of this crop must be supplied from a relatively small soil volume, requiring intensive cultural management. In addition, onions are physiologically sensitive to environmental influences. Day length and temperature are particularly important in determining the timing and degree of vernalization, bulbing, and other processes. Interactions of moisture and fertility with these environmental conditions significantly affect plant growth responses. As a result, onions are quite sensitive to deficiencies of a number of nutrients.

Onion production is widely distributed from the tropics to northern temperate zones. Within this range are a great diversity of soils ranging from highly acid to alkaline. Nutritional disorders of onions that are of major importance in some areas are unknown in others. Particularly significant is the large commercial produc-tion on organic soils in Europe and North America. Whereas most mineral soils supply adequate amounts of the micronutrient elements, organic soils tend to bind these micronutrients, leading to serious deficiencies in the plants.

Compared with other vegetables, an onion crop removes moderate amounts of nutrients from the soil. These nutrients, however, must be absorbed from a smaller soil volume than that of crops with a more exten-sive root system. Because onions have a high harvest index, nutrient removal varies directly with yield. Table 12.1 contains representative nutrient removal data at different yields for onions grown in California.

Deficiency Symptoms

Nitrogen (N)

Nitrogen status is the major nutritional concern worldwide in onion production on mineral soils. The mobility of nitrate in both soil and the plant makes careful management necessary. Deficiencies of N can lead to small plants and bulbs and early maturity. Conversely, excess N produces soft bulbs, increased susceptibility to field and storage rots, and delayed maturity.

Table 12.1. Nutrient removal by bulb and green bunching onion crops[a]

Yield		N		P		K		Ca		Mg		Na	
cwt/acre	t/ha	lb/acre	kg/ha	lb/acre	kg/ha	lb/acre	kg/ha	lb/acre	kg/ha	lb/acre	kg/ha	lb/acre	kg/ha
Bulb onions													
313	35	76	85	15	17	56	63	23	26	4.1	4.6	1.5	1.7
406	45	114	128	21	24	88	99	25	28	5.6	6.3	3.1	3.5
479	54	150	168	26	29	80	90	45	50	7.8	8.7	4.7	5.3
495	55	132	148	23	26	98	110	36	40	7.3	8.2	3.2	3.6
538	60	133	149	26	29	99	111	45	50	8.0	9.0	2.4	2.7
Green bunch-ing onions													
...	...	66	74	9	10	79	88	28	31	7	8	8	9

[a]Data from Zink (14,15).

Nitrogen deficiency is characterized by uniformly light green foliage that becomes increasingly yellow as the deficiency progresses (Plate 165). Labile chlorophyll molecules are broken down, and the nitrogenous compounds are translocated to the other parts of the plant. Thus symptoms are first visible in the older outer leaves and move progressively toward the newly emerging leaves in the center of the plant as N is translocated toward the stronger sinks near the growing point. Leaf tips die back until whole leaves are lost. Nitrogen-deficient onion plants tend to have thin, small-diameter, erect leaves. Plants grow slowly, often becoming stunted.

Low N at the time of seedstalk initiation may increase the number of seedheads formed and thus increase seed yield. High N keeps the bulbs vegetative and reduces both the number of seedstalks and total seed yield.

Phosphorus (P)

Phosphorus deficiency in onions has been associated with slow plant growth (Plate 166), delayed maturity, and a high percentage of thick necks at harvest. Because P is essential for energy transfer within the plant, all aspects of growth are affected. Deficient plants are initially recognized by the dull green color of the leaves. Leaf tips then wilt and eventually die without the yellowing associated with N and K deficiency. The necrosis advances toward the base of the leaves, sometimes leaving islands of green within the yellow or brown necrotic tissue. In the final stage, the dead tissues turn black. A significant amount of P in the plant occurs in the phospholipids that compose plant membranes. Thus P deficiency affects the transport of other elements within the plant, especially divalent cations such as Ca. Imbalance of P may be responsible for observed deficiency symptoms of other elements.

Potassium (K)

Onions are fairly susceptible to K deficiency. Insufficient K, which acts as a cell buffer and an enzyme activator, produces a variety of symptoms. Leaves of deficient plants initially become dark green from accumulation of nonstructural carbohydrates caused by reduced protein synthesis. Later, the tips of the oldest leaves begin to wilt, especially on the upper sides. Leaves then droop, though remaining somewhat inflated, and the wilted areas progress from satiny to a wrinkled, crepe-paper-like appearance. As the necrosis progresses, leaf coloration becomes similar to that seen in severe N deficiency.

Calcium (Ca)

Calcium deficiency of onions is seldom, if ever, encountered under field conditions. When grown with Ca-free nutrient solutions in the greenhouse, young leaves become pale, die back, and eventually die. In some plants an area of the leaf may die, causing the distal section to fall over. Lack of Ca in young plants may also cause the root system to collapse, but if Ca is supplied during early growth and then removed, leaf tips die back without any significant growth restriction.

Calcium is important to the formation and stability of membranes. Insoluble Ca pectates in the middle lamella are also essential for holding adjacent cells together. Some symptoms of Ca deficiency can be directly attributed to structural failures in the cell wall. Because Ca is usually metabolized into these immobile forms, little is redistributed within the plant, and most deficiencies occur in the most actively growing tissues.

Magnesium (Mg)

Magnesium deficiency in onion is generally characterized by slow plant growth and even death of weak plants. Older leaves become uniformly yellow along the entire length due to loss of chlorophyll, of which Mg is a key element. Chlorophyll may fail to form in younger leaves. In chronic deficiencies, chlorosis is followed by tip browning and dieback. The interveinal chlorosis characteristic of Mg deficiency in many plants is not readily apparent in onion leaves.

Sulfur (S)

Sulfur deficiency is seldom encountered in the onion-growing regions of the United States. Naturally occurring organic and inorganic S and additions from atmospheric pollution generally supply the needs of the plants. However, onions are somewhat sensitive to low S. Deficient plants tend to produce fewer leaves, but bulb development is usually normal. New leaves exhibit a uniform yellowing caused by a lack of chlorophyll-binding proteins, and may in extreme cases become thick and deformed.

Sulfur compounds play a major role in determining the taste and pungency of onion bulbs. Sulfur-propylene derivatives of cysteine and cysteine sulfoxide are major constituents of onion and garlic. Sulfur-containing glycosides and volatile compounds provide the characteristic flavor and aroma. Reducing S inputs may lower pungency and produce a more palatable onion.

Iron (Fe)

Onions appear to be insensitive to Fe deficiency in the field. In laboratory studies, tissue Fe levels and growth of onions receiving no Fe were not significantly different from those with adequate Fe, and no deficiency symptoms were observed.

Zinc (Zn)

Onions are very sensitive to Zn deficiencies, which occur in soils with high pH or P and in situations where calcareous subsoils are exposed by deep tillage or land leveling. NADH dehydrogenase and other chloroplast enzymes are Zn-dependent, so the primary effect of a deficiency is on photosynthesis. Zinc-deficient onion plants are stunted, with marked twisting and yellow striping of the leaves (Plate 167). Cool, wet weather increases the severity of the deficiency. High P restricts Zn transport within the plant and may aggravate the problem.

Manganese (Mn)

Onions are one of the vegetable crops most susceptible to Mn deficiency. Failure of photosystem II reaction centers results in leaves that are initially light colored (Plate 168). The outer leaves show striped interveinal chlorosis, and these leaves develop tipburn and sometimes curl, with development of a progressive necrosis. Plants may become seriously stunted, and bulbing is often delayed. At maturity, a high percentage of bulbs have thick necks. Drought has been shown to increase the severity of the symptoms. Because soil Mn levels often exhibit a high spatial variability, severe deficiency symptoms may disappear as the root system expands and develops into soil with a higher Mn level.

Copper (Cu)

Copper is usually deficient only in organic soils that have had no Cu fertilization. When severe, Cu deficiency may cause onion leaves to turn white and twist into spirals (pigtail) or bend at right angles to the plant (Plate 169). Leaf tips may become chlorotic and then die back. Bulbs are less solid than those with adequate Cu and tend to mature earlier.

Bulb quality has been shown to be significantly reduced by Cu deficiency. Outer bulb scales are thin and poorly pigmented, resulting in pale yellow bulbs (Plate 170). Most peat and muck soils used for onion production have been fertilized with Cu and do not respond to further Cu applications because Cu is not leached or removed in large quantities.

Boron (B)

Onions are less responsive to B than to many other micronutrients. However, severe B deficiency can cause obvious symptoms. Leaves tend to become stiff, brittle, and dark green. Young leaves may be mottled yellow and green. Older leaves often become chlorotic, with tip dieback and distorted, shrunken areas. Regularly spaced transverse yellow lines near the base of the leaves may develop into a series of ladderlike cracks. A few isolated cracks may also occur 1- to 1½-in. (3- to 4-cm) from the leaf tip.

Molybdenum (Mo)

Onions are highly sensitive to Mo deficiency, which is most often associated with acidic peats and sandy soils. Poor emergence and death of seedlings may occur in highly deficient soils. As the plant grows, leaf tips die back with a 1- to 1½-in. (3- to 4-cm) zone of flabby, wilted tissue between the necrotic and healthy tissues (Plate 171). This wilting and necrosis progress down the leaf and in severe cases may kill the plant. Because

Mo functions primarily in nitrate reductase, these symptoms in deficient plants result from reduced conversion of nitrate N to ammonium forms (Plate 172).

Toxicity Symptoms

Toxicity symptoms directly attributable to naturally occurring elements are seldom seen in onions under field conditions. Most plant injury from mineral elements is either from excess salt accumulation or from introduction of unusually high concentrations of elements that are normally absent or insignificant in agricultural soils.

Salinity

Salt toxicity is a potential problem in many irrigated onion production areas. Upward capillary water movement in the soil in response to evapotranspiration may result in accumulation of salts in the root zone. These problems are especially serious where highly saline water is used for irrigation. Uptake of these salts may result in elevated tissue levels of sodium (Na) and chlorine (Cl). However, the dull leaf color and tipburn associated with onions grown under these conditions are primarily symptoms of induced physiological drought rather than toxicity of a given element.

Chlorine

Onions have been found to be quite susceptible to Cl toxicity. This disorder is usually associated with irrigation with Cl-containing water. Chlorine toxicity is manifested by chlorosis and tipburn and stunting of the plants.

Critical and Sufficient Nutrient Tissue Levels

Nutrient imbalance in onions, as in many horticultural crops, may reduce quality without affecting total yield. Economic return to onion growers may be completely lost to poor quality in a crop with average to good yields. Quality factors such as bulb firmness and color are determined by the interaction of nutrition and genetics with yield, management practices, and the environment. In addition, tissue elemental content varies with plant age, both in the total plant and in specific plant tissues.

Because of the complexity of these relationships, specific critical values for most of the nutrient elements have not been established for onions as they have been for many agronomic crops. Some published deficiency, sufficiency, and toxicity levels for onions are listed in Table 12.2. Often these sufficiency levels are lower than reported critical values for many other vegetable crops. This observation is supported by sequential tissue sampling studies comparing a number of vegetables. Over the growing season, levels of N, P, and K in onion tissue were substantially lower than in other vegetables, and the decline in P with time was much more rapid than for any other vegetable tested.

Onion plants produce most of their green leaf area during the vegetative growth stage, with little further leaf development once bulbing is initiated. The stem plate continues to initiate leaves, but these are modified to form the bulb. As the bulb grows, mobile elements are often translocated from the older photosynthetic leaves to the newly expanding bulb scales. Consequently, sampling leaf tissue as an indication of nutrient status may produce an inconsistent or inaccurate picture. Similarly, deficiency symptoms appearing late in the crop season must be considered in relationship to this redistribution of elements within the plant. Root tissue nitrate has been shown to correlate more closely than leaf nitrate with soil nitrate levels and plant growth. It is possible that roots may be the plant part of choice for determining the status of other nutrients as well.

Tissue and soil levels of mobile elements are also sensitive to irrigation regimes and precipitation, particularly under raised bed cultural systems. Sprinkler

irrigation or rainfall may move these elements below the reach of the limited root system. Furrow irrigation, on the other hand, tends to move fertilizer salts toward the top of the beds. Curves generated from weekly soil and tissue sampling often show variations that correspond to additions of water. Changes in plant growth rate due to other environmental conditions such as temperature and available radiation may interact with nutrient availability in the root zone to significantly affect tissue nutrient levels. Therefore, tissue and soil sampling should be standardized as much as possible on soil moisture, and careful consideration should be given to environmental and cultural conditions when interpreting the results. Similarly, nutrient deficiency symptoms in the leaves sometimes reflect temporarily low levels of a given nutrient that will self-correct as environmental conditions change.

Management Considerations

Careful planning to avoid fertility problems can do much to prevent nutrient deficiencies in onions. Because onions are grown in relatively small blocks in comparison to field crops, it is often possible to avoid areas with localized fertility problems. For example, exposed calcareous subsoils or very acid peats may reduce availability of certain elements and should be avoided when possible. Correction of extreme soil pH can allow onion production in many soils. Onions generally grow best from pH 6.5 to 7.5. At pH values much above 8.0 or below 6.0, growth and quality may be significantly reduced. Liming is widely practiced to bring acid peats and mineral soils to an acceptable pH.

Most of the major elements are supplied to the onion crop from fertilizers, and general fertilizer recommendations are available for all major production areas. Actual recommendations vary widely because of differences in environmental conditions and soils. In addition, the growing season in the field for bulb onions of different maturity classes and day length requirements ranges from 4 to 6 months or more. Obviously, no one set of recommendations can maximize production in all situations. Blind adoption of cultural systems from other areas may lead to serious nutrient imbalances.

In most production systems, most fertilizer nutrients are applied preplant and incorporated into the bed. However, N side-dressings often are necessary for optimum production in many soils where N is leached away from the limited root system. Careful irrigation management can reduce N losses in many systems. Differences in wetting patterns caused by the type of irrigation system should be considered in fertility management.

Micronutrient deficiencies are best corrected before planting the crop. Small amounts of these elements are needed to correct many deficiencies, and uniform

Table 12.2. Representative published tissue levels of nutrients in onions

Element	Deficient	Sufficient	Toxic
N, %	2–2.5	2.5–3	...
	0–2	1.5–2.5	
		2.5	
P, %	0.10	0.20	...
	0–0.1	0.26	
		0.25–0.40	
		0.2	
K, %	2–34	3–4.5	...
	2	2.5	
	0.8	4.1	
Ca, %	0.18	0.52	...
		1.6–1.28	
		1.5–3.5	
Mg, %	0.034	0.37	...
Mn, ppm	0–10	16–24	...
		20	
		55–65	
B, ppm	...	10	91–1,578
		27–33	58–1,186
		29–44	
		30–45	
Zn, ppm	0.5	22–32	39
		10–15	30
		20	
Fe, ppm	...	29–50	780
Al, ppm	...	63	1,500
Cl, %	...	0.25	0.53–1.53
		4–5	
		0.32–0.42	
Co, ppm	...	0.02–0.13	...

distribution of the fertilizer is necessary to prevent wide spatial variability in nutrient levels. Because some micronutrients are relatively immobile in the soil and are removed in small amounts, applications may be necessary only at intervals of some years. For example, onions grown on some muck soils on which Cu deficiencies were corrected showed no symptoms of Cu deficiency for many years even when no additional Cu was applied.

The use of foliar fertilizer applications to prevent or overcome nutrient deficiencies has been promoted for a number of crops including onions. Few published research data are available to recommend this approach in onion production. In addition to the obstacle of developing highly soluble forms of fertilizer that can be applied in sufficient concentrations to significantly alter plant growth, the vertical orientation and waxy cuticle of onion leaves make penetration of these solutions very difficult. Most of the solution thus falls to the soil, so any plant response is likely to be as much from root uptake as from foliar absorption.

Selected References

1. Brewster, J. L. 1977. The physiology of the onion. Hortic. Abstr. 47:17-23, 103-112.
2. Chapman, H. D. 1966. Diagnostic Criteria for Plants and Soils. Division of Agricultural Sciences, University of California, Riverside.
3. Geraldson, C. M., Klacan, G. R., and Lorenz, O. A. 1973. Plant analysis as an aid in fertilizing vegetable crops. Pages 365-378 in: Soil Testing and Plant Analysis. L. M. Walsh and J. D. Beaton, eds. Soil Science Society of America, Madison, WI.
4. Greenwood, D. J., Barnes, A., Liu, K., Hunt, J., Cleaver T. J., and Loquens, S. M. H. 1980. Relationships between the critical concentrations of nitrogen, phosphorus and potassium in 17 different vegetable crops and duration of growth. J. Sci. Food Agric. 31:1343-1353.
5. Knott, J. E. 1933. The effect of certain mineral elements on the color and thickness of onion scales. Cornell Agric. Exp. Stn. Bull. 552.
6. Maynard, D. N., and Lorenz, O. A. 1989. Onions. Pages 189-199 in: Detecting Mineral Nutrient Deficiencies in Tropical and Temperate crops. D. L. Plucknett and H. B. Sprague, eds. Westview Press, Boulder, CO.
7. Purvis, E. R., and Carolus, R. L. 1964. Nutrient deficiencies in vegetable crops. Pages 245-286 in: Hunger Signs in Crops: A Symposium. H. B. Sprague, ed. David McKay, New York.
8. Scaife, A., and Turner, M. 1984. Diagnosis of mineral disorders in plants. II. Vegetables. Chemical Publishing, New York.
9. Stuart, N. W., and Griffin, D. M. 1944. Some nutrient deficiency effects in the onion. Herbertia 2:329-337.
10. Stuart, N. W., and Griffin, D. M. 1946. The influence of nitrogen nutrition on onion seed production in the greenhouse. Proc. Am. Soc. Hortic. Sci. 48:398-402.
11. Tyler, K., May, D., Guerard, J., Ririe, D., and Hatakeda, J. 1987. Dehydrator onion mineral nutrition. Univ. Calif. Veg. Crops Ser. 223.
12. Vitosh, M. L., Warncke, D. D., and Lucas, R. E. 1973. Secondary- and micro-nutrients for vegetables and field crops. Mich. State Univ. Ext. Bull. E-486.
13. Wannamaker, M. J., and Pike, L. M. 1987. Onion responses to various salinity levels. J. Am. Soc. Hortic. Sci. 112:49-52.
14. Zink, F. W. 1962. Growth and nutrient absorption of green bunching onions. Proc. Am. Soc. Hortic. Sci. 80:430-435.
15. Zink, F. W. 1966. Studies on the growth rate and nutrient absorption of onion. Hilgardia 37(8):203-218.

Chapter 13

Gerald E. Wilcox
Horticulture Department
Purdue University, West Lafayette, Indiana

Tomato

Tomato is a herbaceous perennial but is generally grown as a frost-tender annual, with optimum growth in the temperature range of 60–90° F (16–32° C). Tomatoes pass through several stages of development in the course of a season's growth—seedling establishment, vegetative growth, flowering, and fruiting—with each stage differing in nutrient requirements.

The period of seedling establishment and early growth is devoted to development of a root system and vegetative top. A seeded tomato develops a strong taproot that grows straight down to a depth of 2–3 ft (60–90 cm). A transplanted tomato develops a secondary branch root system. If the taproot encounters a hardpan, root system development is restricted to the zone above the compacted layer, which can severely limit the soil volume it explores for water and nutrient uptake.

Vegetative development alone occurs during the first 42 days, after which fruit develops concurrently (Fig. 13.1). Growth is rapid during the next 4 weeks while the plant is flowering and setting fruit. After 70 days, there is very little further development in the vegetative portion of the plant, and dry matter accumulates primarily in the fruit at a fairly constant rate. Fruit ripening begins at 84 days and progresses to harvest at 112 days.

Nutrient uptake is continuous during growth and development of the tomato plant. If the nutrient availability in the soil is high, the plant's needs can be met through root absorption during the entire growing period. However, if soil nutrient availability is limiting, the requirement for fruit development will be a function of translocation from the vegetative portions in combination with root uptake. Limited uptake of Ca due to drought causes a deficiency that cannot be compensated for by translocation, resulting in blossom-end rot in fruit.

Deficiency Symptoms

Nitrogen (N)

Nitrogen-deficient plants grow slowly. Leaves are small and light green to yellowish green to pale yellow (Plate 173). Leaves near the top will be yellow-green with purple veins. Stems are thick and hard. Flower buds turn yellow and drop off. Fruits may be small and pale green before ripening. Yields are reduced.

Phosphorus (P)

Plants deficient in P grow slowly, and maturity is delayed. Seedling growth is stunted, especially during cool weather. Leaves become dark green with purple interveinal tissue on the underside of the leaf (Plate 174). Stems become slender, fibrous, and hard. Seedlings require a high concentration of P for normal growth. Transplants treated with a solution of 3 lb of 10:52:17

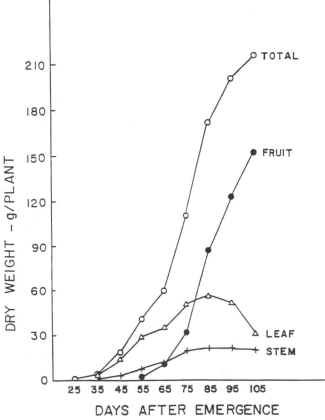

Fig. 13.1. Dry weight accumulation in the tomato plant over a 105-day growing period.

fertilizer (or the equivalent soluble-P salt) per 50 gal of water (1.4 kg/190 L) at one cup (250 ml) per plant make optimum recovery after transplanting. For direct-seeded tomatoes, the application of 1 gal of 10:34:0 per 1,000 ft of row (3.8 L/305 m), 1–2 in. (2.5–5.0 cm) directly under the seed is recommended. For applications directly on the seed, dilute 10:34:0 at a 1:5 ratio with water and apply 1 pint of solution per 100 ft of row (0.5 L/30 m).

Potassium (K)

Young plants have dark green leaves with small stems and shortened internodes. In a field of K-deficient plants, the plant tips have a spikelike appearance resulting from the slower development of the leaf petioles. Young leaves are dark green, becoming crinkled and curled (Plate 175). Older leaves are chlorotic and bronzed; leaf margins become brown, and tissues break down between the veins (Plate 176). Fruits of deficient plants drop off soon after ripening. Sepals and stems die between the fruit and abscission layer of the pedicel. Fruits of deficient plants are not fleshy, ripen unevenly, and appear blotchy (Plate 177).

Calcium (Ca)

A lack of Ca results in abnormal development of new cells. Tomato seedlings show upward cupping of leaves with necrotic margins (Plate 178). Plants are stunted as tip growth is reduced. Application of KNO_3 intensifies Ca deficiency of seedlings, with almost no growth on soil with less than 0.5 meq of Ca + Mg per 100 g. Root tips die, and roots branch behind the dead tip (Plate 179). Blossom-end rot of tomato fruits is a Ca deficiency, usually induced by water stress in the plant. At time of fruit set, cells at the blossom end of fruits are injured when insufficient Ca translocation to the flower results in a dry-rot brown area on the expanding fruit (Plate 180).

Magnesium (Mg)

Lower and older leaves are first affected by Mg deficiency. Leaf veins remain dark green, and areas between the veins become yellow (Plate 181). Nitrogen deficiency intensifies the development of Mg deficiency on low-Mg soils.

Iron (Fe)

Young leaves are chlorotic in Fe-deficient plants (Plate 182). Pale yellow mottling starts at the base of the leaves and spreads upward along the midribs and outward along the veins. Iron deficiency is often related to soils with high pH, free $CaCO_3$, high P, and poor aeration.

Zinc (Zn)

Young leaves of Zn-deficient plants are small with yellow interveinal mottling. Necrotic interveinal areas occur in expanded and older leaves (Plate 183).

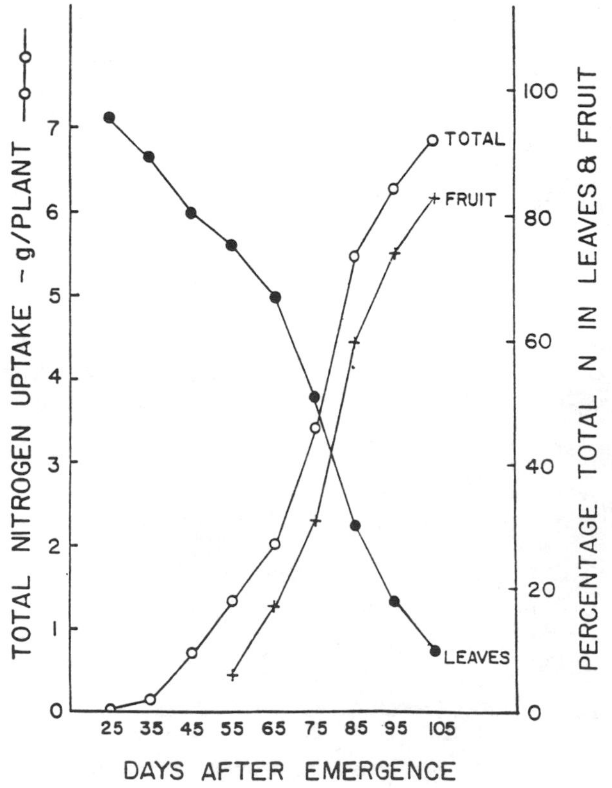

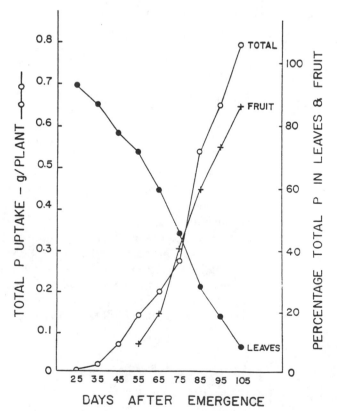

Fig. 13.2. Nitrogen accumulation in the tomato plant over a 105-day growing period.

Fig. 13.3. Phosphorus accumulation in the tomato plant over a 105-day growing period.

Manganese (Mn)

Young leaves show interveinal chlorosis with prominent green veination (Plate 184). Severe deficiency results in necrosis of interveinal tissue. Plants recover rapidly after foliar application of Mn. If deficiency occurs, spray foliage with a solution of 4 lb of $MnSO_4$/ 100 gal of water (1.8 kg/380 L). Apply at a rate of 50–75 gal/acre (75–115 L/ha). In starter fertilizer, band 1 lb of $MnSO_4$/1,000 ft (0.5 kg/300 m) of row.

Boron (B)

Growth reduction from B deficiency affects the growing tip. The terminal shoot turns inward and dies. Plants with B deficiency have small, crinkled, deformed leaves with large irregular areas of discoloration. Tomatoes require more B than beans but much less than beets or cabbage. Boron deficiency is usually caused by over-liming. Lime application to acid sands to raise pH to near neutral may cause B deficiency (Plate 185).

Molybdenum (Mo)

Molybdenum-deficient leaves show a characteristic cupping or curling in the margins, yellowing between the veins, and death of the growing tip.

Toxicity Symptoms

Aluminum (Al)

Aluminum toxicity occurs on acid soils of pH 5.0 or less. Excess Al results in stunted root and top growth. The underside of older leaves turns purple (Plate 186). The seedling also develops a white interveinal chlorosis in the older leaves.

Boron

Excess B exuded from hydathodes of older leaves causes marginal leaf burn (Plate 187). Tomatoes are semitolerant of B toxicity.

Manganese

Toxicity symptoms include reduced growth rate and necrosis along the main veins (Plate 188). In plants grown in soil, toxicity symptoms start on the lower leaves and work up the main stem. The leaves (blades and petioles) die back to the stem and hang on by mechanical attachment. Necrotic streaking develops on the main stem. Necrotic tissue along the main vein of the leaf blades is bordered by yellow chlorotic areas.

Zinc

Zinc toxicity causes a marked shortening of the internodes of plants that otherwise appear normal (Plate 189).

Ammonia

Tomato seedlings are sensitive to ammonia (NH_3), which causes chlorosis in the intervcinal areas of the developed leaves (Plate 190). The leaflets develop an upward roll. Toxicity can result from use of diammonium

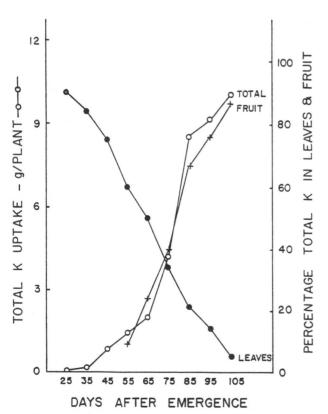

Fig. 13.4. Potassium accumulation in the tomato plant over a 105-day growing period.

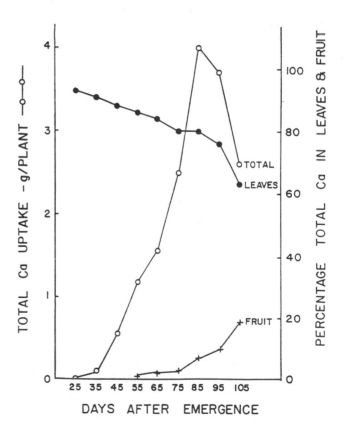

Fig. 13.5. Calcium accumulation in the tomato plant over a 105-day growing period.

phosphate fertilizer, which increases the ammonium concentration above a critical level. With nitrification—conversion of the ammonium to nitrate—the problem soon disappears.

Disorders Resembling Deficiencies or Toxicities

Seeded Tomatoes, Uneven Emergence

Tomato seeds germinate and push up through the soil cover. If the soil above the seed has crusted, emergence can be very uneven because of differential impedance by the compacted soil. Soil compaction can also affect the growth rate of the tomato seedling after emergence. A dense soil will seriously reduce growth and development of the tomato plant.

Effects of Cold

At air temperatures below 58°F (15°C), growth is slowed. The undersides of leaves can turn purple, resembling P deficiency.

Plant Growth and Nutrient Uptake

Total dry weight is partitioned between leaves and stems up to 55 days after emergence, at which time fruit begins to develop (Fig. 13.1). Vegetative increase in dry weight is rapid and continues in stems and leaves to the 70th day, with very little increase after 77 days. Dry

Table 13.1. Nutrient composition of expanded tomato leaf during fruit accumulation period

Nutrient	Concentration[a]	
	Deficient	Adequate
N, %	<2.5	>3.5
P, %	<0.12	>0.2
K, %	<2.3	>3.5
Ca, %	<1.0	>3.0
Mg, %	<0.3	>0.42
S, %	<0.15	>0.25
Fe, ppm	<50	>100
Zn, ppm	<10	>15
Mn, ppm	<24	>35
Cu, ppm	<4	>8
B, ppm	<15	>25
Mo, ppm	<0.1	>0.3

[a]Concentrations that fall between adequate and deficient nutrient levels may induce a response in plants.

Table 13.2. Composition of tomato plant parts in relation to stage of growth

Days after emergence	Plant part	Nutrient concentration (% dry wt)				
		N	P	K	Ca	Mg
21	Whole plant	4.38	0.54	4.35	2.58	0.73
28	Whole plant	4.15	0.45	4.20	2.92	0.79
35	Whole plant	3.72	0.36	4.17	2.63	0.74
42	Whole plant	4.30	0.42	4.45	2.59	0.74
49	Branch	2.84	0.36	4.94	1.22	0.57
	Leaf	3.99	0.35	4.14	2.95	0.80
	Fruit	3.10	0.53	3.95	0.12	0.19
56	Branch	2.74	0.35	5.21	1.40	0.65
	Leaf	4.09	0.34	4.37	3.08	0.84
	Fruit	2.95	0.47	4.44	0.12	0.23
63	Branch	2.64	0.41	4.53	1.13	0.59
	Leaf	3.96	0.34	3.76	3.06	0.84
	Fruit	3.40	0.56	4.97	0.14	0.24
70	Branch	2.48	0.33	4.30	1.07	0.58
	Leaf	3.75	0.31	4.33	3.09	0.92
	Fruit	2.94	0.50	4.56	0.30	0.26
77	Branch	1.85	0.24	3.77	1.25	0.64
	Leaf	3.34	0.26	3.16	4.31	1.15
	Fruit	2.68	0.44	4.47	0.23	0.23
84	Branch	1.88	0.23	3.73	1.38	0.60
	Leaf	3.54	0.25	3.51	3.49	1.00
	Fruit	2.44	0.38	4.38	0.16	0.20
91	Branch	1.58	0.18	3.22	1.63	0.57
	Leaf	2.86	0.22	2.57	4.22	1.10
	Fruit	2.76	0.42	4.62	0.17	0.21
105	Whole plant	1.95	0.20	2.81	3.28	0.80
	Fruit	3.08	0.45	4.88	0.18	0.22

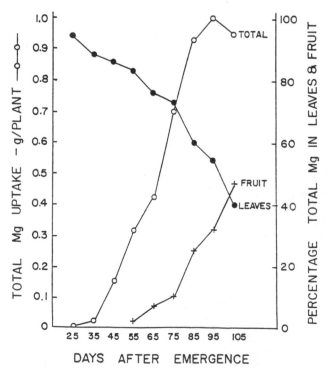

Fig. 13.6. Magnesium accumulation in the tomato plant over a 105-day growing period.

Table 13.3. Tomato tissue composition at various growth stages corresponding to adequate and deficient levels

Medium	Growth stage	Plant part	Level	Nutrient concentration (% dry wt)				
				N	P	K	Ca	Mg
Soil	Seedling (24–31 days after emergence)	Whole plant top	Adequate	3.5	0.35	4.0	...	...
			Deficient	2.8	0.28	3.0	...	...
	Flower and fruit formation (56–77 days)	Leaf blade and petiole	Adequate	3.0	0.35	4.0	...	...
			Deficient	2.0	0.20	2.5	...	...
	First ripe fruit	Leaf blade and petiole	Adequate	2.5	0.25	3.0	...	...
			Deficient	1.5	0.15	1.5	...	...
Solution A	Seedling (30 days after germination)	Whole plant top	Adequate	5.2	0.8	6.3	4.2	0.66
B	Seedling (30 days after germination)	Whole plant top	Deficient	1.5	0.8	3.0	1.5	0.41

weight accumulation in the fruit is nearly linear from day 70 to 105. In a plant population of 10,000 plants/acre (25,000 plants/ha), this amounts to an accumulation of 1,678 lb of fresh fruit/acre (1,880 kg/ha) per day. After the 70th day, almost all dry matter accumulation is in the fruit. On a calender basis in Indiana (planting date May 2), vegetative development occurs until July 10 and fruit accumulation from July 10 to August 10. From August 5 to harvest, fruit weight increases rapidly.

Total N uptake by the tomato crop is 176 lb/acre (197 kg/ha). The N accumulation rate is highest during the rapid vegetative development period, at 0.65 oz (185 mg) per plant per day compared with 0.4 oz (118 mg) per plant per day during the fruit expansion period (Fig. 13.2).

The percentage of total N in the leaves drops steadily from the seedling stage. About 80% of the total N accumulated is in the leaves at the start of fruit accumulation (day 70). By the 84th day only 33% of the total N accumulated to that time is in the leaves, and about 52% is in the fruit. At harvest, about 24% of the total N is in the leaves and about 68% is in the fruit (Fig. 13.2).

The total P uptake in the tomato crop is 23 lb/acre (26 kg/ha) or 0.002 lb/plant (1.052 g), of which 75% is in the fruit and about 18% in the leaves. The total P uptake is particularly rapid during the fruit accumulation period (Fig. 13.3).

The total K uptake in the tomato plant is 0.026 lb/plant (12 g). At harvest nearly 70% is in the fruit and 16% in the leaves. The total K uptake is 264 lb/acre (296 kg/ha), with about 185 lb (84 kg) in the fruit. The K uptake is rapid and fairly constant during vegetative and fruit expansion, 4.5 lb/acre (5 kg/ha) per day from day 49 through day 105 (Fig. 13.4).

Table 13.4. Tomato Mg composition related to soil Mg levels

Soil exchangeable Mg		Mg concentration (% dry wt)		
lb/acre	kg/ha	Base leaf	Recently mature leaf	Expanding leaf
32	36	0.27[a]	0.35	0.42
52	58	0.42	0.46	0.45
63	71	0.57	0.53	0.49
76	85	0.70	0.66	0.60

[a]Deficiency symptoms are visible.

The total Ca uptake is about 100 lb/acre (112 kg/ha) by the end of the fruit accumulation period. The Ca is partitioned at 77% in the leaves and 7% in the fruit at harvest (Fig. 13.5).

Total Mg uptake is 31 lb/acre (35 kg/ha). Uptake is constant from first bloom to fruit maturation at 95 days. At harvest, 58% of the Mg is in the leaves and 27% in the fruit (Fig. 13.6).

Nutrient levels in the tomato plant can be a diagnostic tool. Diagnosis of plant nutrition can be accomplished with a knowledge of the composition of plant parts that relates to adequate levels. In fact, the plant can be deficient in an element that affects growth rate before other visible symptoms appear. The nutrient concentrations in plant parts relating to adequacy and deficiency are given in Tables 13.1–13.4; these can be used for reference when tissue analysis is used to diagnose the nutrient status of the plant.

Selected Reference

1. Jones, J. B., Jones, J. P., Stall, R. E., and Zitter, T. A., eds. 1991. Compendium of Tomato Diseases. American Phytopathological Society, St. Paul, MN.

Chapter 14

R. Hall
Department of Environmental Biology
University of Guelph, Guelph, Ontario

H. F. Schwartz
Department of Environmental Biology
University of Guelph, Guelph, Ontario

Common Bean

The common bean, *Phaseolus vulgaris* L., is grown extensively in eastern Africa, North and Central America, South America, eastern Asia, and western and southeastern Europe. It is a member of the Leguminosae, tribe Phaseolae, subfamily Papilionoideae. In growth habit, the plants may be determinate, where reproductive terminals develop on the main stem and no further nodes are produced on the main stem after flowering, or indeterminate, in which case a vegetative terminal occurs on the main stem and flowering does not lead to cessation of vegetative growth. Germination is epigeal, and time to flowering is 28–42 days. The plant produces a taproot and a fibrous root system and fixes nitrogen in root nodules developed in response to colonization of the root by the bacterium *Rhizobium phaseoli*. As many as two-thirds of the flowers may abscise, and some fruits and seeds may abort. Physiological maturity of the seed may occur 60–150 days after planting. The green pods in snap beans or dry seeds in dry beans are consumed as food.

Deficiency Symptoms

Nitrogen (N)

Beans are legumes and can fix N in the presence of appropriate strains of *Rhizobium* unless cultural, varietal, or inoculation difficulties limit this fixation ability and leave the plant dependent on residual soil nitrogen or applied N fertilizer. Nitrogen deficiency can occur on all soils and is especially severe in sandy soils with low organic matter. It may be induced by heavy rain or excess irrigation via leaching of nitrates or by suppression by anaerobiosis of N-fixing bacteria. Soil pH outside the range 6.0–8.0 may restrict the availability of the element. Nitrogen is an essential component of amino acids, proteins, chlorophyll, nucleotides, and nucleic acids.

Deficiency symptoms appear as a uniformly pale green to yellow discoloration of older leaves. Growth is reduced, and few flowers develop or pods fill poorly.

Nitrogen deficiency can be corrected by application of N fertilizer and organic matter. The amount of N required may range from 18 to 90 lb of N/acre (20 to 100 kg of N/ha), depending on soil, crop, and other factors. Soil inoculation with appropriate strains of *Rhizobium* can be successful but is affected by specificity between the *Rhizobium* strain and its host cultivar, interaction with native and less efficient rhizobia, soil pH, soil temperature, soil N and P contents, pesticide use, and farming practices.

Phosphorus (P)

Phosphorus deficiency can occur on many soils, especially those with low pH or that are highly leached. Its availability becomes markedly restricted below pH 6.2. Phosphorus is part of nucleotides, nucleic acids, phosphorylated sugars, cell membranes, and some coenzymes. Phosphates act as buffers to help maintain cell pH.

Slow growth is the main symptom of P deficiency. Upper leaves are small and dark green, and older leaves may turn yellow, then brown, and senesce prematurely. Plants are often stunted, and they have thin stems and shortened internodes. The vegetative period may be prolonged, and the flowering phase is delayed and shortened. The number of aborted flowers is often high, few pods form, and seed quantity may be reduced. Phosphorus deficiency can be controlled chemically by band application of various rock phosphates or superphosphate fertilizers. Some cultivars are more tolerant of low P levels than others.

Potassium (K)

Potassium deficiency may occur in soils with low fertility, high calcium and magnesium contents, low pH, or a highly permeable sandy texture. Potassium becomes less available below pH 6.0. The element is important in the maintenance of cell turgor and in movement of stomata. It activates many enzymes and appears to assist in the translocation of anions from one plant part to another.

Deficiency symptoms usually appear in young plants. They are seen first as a marginal chlorosis of older leaves, followed by a marginal scorch (which may resemble common bacterial blight, but without the water-soaking) between the veins. The yellowing spreads towards the center and base of the blade. The leaf may curl downward, but scorched margins will curve upwards. Plants may have weak stems, short internodes, and reduced root growth and can collapse easily. Potassium deficiency can be corrected by application of potassium chloride or potassium sulfate. Some cultivars of beans are able to efficiently utilize small amounts of K in the soil.

Calcium (Ca)

Acid soils with a pH between 4 and 5.5 normally have low levels of Ca and Mg. The maximum availability of Ca occurs in the pH range 7.0–8.5. Calcium deficiency often occurs together with Al and Mn toxicities. Calcium is part of the intercellular "cement," activates certain enzymes, influences the semipermeability of cell membranes, and may confer stability on the mitotic spindle apparatus and detoxify oxalic acid.

Symptoms of Ca deficiency include loss of turgor, death of growing points, dark green older leaves, yellowing of newer leaves, and sometimes hypocotyl collapse. Older leaves eventually scorch and senesce. Pods may also be soft and yellowed, and seeds may fail to develop. Plant height and dry matter production are often reduced. Calcium deficiency may be corrected by liming and application of superphosphate.

Magnesium (Mg)

Magnesium deficiency generally occurs in acidic soils with low base levels and on volcanic ash soils with low levels of K and Ca. It becomes progressively less available below pH 6.5. As part of the chlorophyll molecule, Mg is essential to photosynthesis. It is also essential to, or enhances, the activity of hundreds of enzymes.

Deficiency symptoms include interveinal chlorosis and rusty speckling to necrosis of the upper surface of lower leaves. These interveinal spots may measure 0.5 mm and are angular and slightly sunken. Magnesium deficiency is normally corrected by the application of dolomitic lime on low-pH soils. On slightly acid or higher-pH soils, magnesium sulfate or potassium-magnesium sulfate is normally used.

Sulfur (S)

Sulfur deficiency seldom occurs in bean production areas anywhere in the world. One incident reported in Michigan occurred on slightly acid soils. The availability of the element declines below pH 6.0. Among its many functions, S is a component of the amino acids cystine, cysteine, and methionine, and thus of most proteins, as well as of thiamine, biotin, and coenzyme A.

Deficiency symptoms are similar to those of N deficiency. A pronounced, uniform, yellow chlorosis of younger leaves develops, and growth ceases. Older leaves may also be chlorotic. A foliar test for nitrate helps identify S deficiency. Sulfur deficiency can generally be corrected by the application of S-containing fertilizers or elemental S.

Iron (Fe)

Iron deficiency can occur in calcareous soils containing free calcium carbonate, high-pH soils, or acidic soils that have been overlimed. Its availability declines above pH 6.0. Excessive P may precipitate the available Fe as insoluble iron phosphate. Temporary deficiencies (less than 48 hr) may occur on younger leaves of bean plants grown on high-pH soils that have been saturated with moisture from rainfall or irrigation, especially if followed by cool day temperatures.

As a component of cytochromes and ferredoxin, Fe is essential to photosynthesis and respiration. It is also part of catalase and peroxidase enzymes that catalyze the conversion of hydrogen peroxide to water and oxygen. It is further found in the enzymes nitrite reductase and nitrate reductase, which are essential in N metabolism.

Symptoms of Fe deficiency appear in young leaves, which become pale yellow, almost white, while the veins remain green (Plates 191 and 192). Profuse, irregular necrosis may develop on severely chlorotic leaves. Fully expanded leaves curve downward, and leaf tips may wilt. Young, unexpanded leaflets may senesce. Iron deficiency can be corrected by application of iron chelates to soil or ferrous sulfate (0.5%) to leaves. Some cultivars are less sensitive to low levels of Fe.

Zinc (Zn)

Zinc is most available in the pH range 5.0–7.0. Zinc deficiency can occur in soils with a high pH, or in acid soils that have received too much lime or P. The problem can also be aggravated by soil compaction, low organic matter, and excess application of manure or crop residue. Elevated absorption of other nutrients, such as Fe, can also induce Zn deficiency. Zinc is essential for the activity of several enzymes such as lactic acid dehydrogenase and alcohol dehydrogenase.

Where Zn is deficient, younger leaves are pale green with yellow tips and margins. They develop an interveinal chlorosis (Plate 193) and become deformed, dwarfed, and crumpled. Older leaves may develop necrotic areas in and between veins. Terminal blossoms and pods may abort. Plants become dwarfed or deformed and are slow to mature. If the deficiency is severe, new leaves are white and plants may die. Deficiencies may appear in spots within a field or throughout the entire field.

Zinc deficiency can be corrected by application of zinc sulfate to soil at the rate of 5–11 lb/acre (6–12 kg/ha) or to leaves as a foliar spray (0.5% zinc sulfate). Zinc chelates applied to soil are also effective. Some cultivars are more sensitive to low levels of Zn than others.

Manganese (Mn)

Manganese deficiency can occur in high-pH, organic, poorly drained, or overlimed soils. The element is most available in the pH range 5.0–6.5. Manganese enhances electron transport from water to chlorophyll and activates many enzymes.

Manganese deficiency is characterized by uniform yellowing of the blade of younger leaves, with the veins remaining green (Plates 194 and 195). The leaves may be finely speckled or may appear pimply when viewed close-up. Older leaves are smoother and generally chlorotic. Pods may be yellow and unfilled. Manganese deficiency can be alleviated by addition of Mn salts in the fertilizer. Band applications of fertilizer may be more effective than broadcast applications. Manganese deficiency detected early in the season can be treated with a foliar spray of $MnSO_4$ at 4 lb/acre (5 kg/ha). Manganese chelates applied to high-pH soil have not been effective.

Copper (Cu)

Copper deficiency, while rarely affecting beans, can occur in organic, sandy, and excessively limed soils. Its availability declines above pH 7.0. Copper is essential to photosynthesis as a component of plastocyanin and to respiration as a part of cytochrome oxidase. It is also an essential component of polyphenol oxidase and ascorbic acid oxidase.

Plants affected by Cu deficiency are stunted and have short internodes. Young leaves are pale to gray or blue-green. Irregular necrotic areas or blotches appear close to veins near the base of the leaflet. Scorching may develop on one side of the leaflet, which then wilts and senesces. Growing tips may die back, and flowering can be suppressed. Copper deficiency can be corrected by applying 4–9 lb/acre (5–10 kg/ha) of copper sulfate to the soil. Minor deficiencies can be corrected by foliar applications of copper sulfate or copper chelate.

Boron (B)

Boron deficiency is rare in beans but can occur in coarse-textured soils having a low organic matter content and high levels of aluminum and iron hydroxide, in alluvial soils with high pH and low total B content, and in neutral or alkaline soils subjected to dryness and high light intensity. Maximum availability of the element occurs at pH levels of 5.0–7.0 and higher than 8.7. Plants require only a fraction of 1 ppm of B, and several parts per million may be toxic.

The functions of B in the plant are not fully known. Boron combines with 6-phosphogluconate to form a 6-phosphogluconate-borate complex. This blocks the pentose shunt pathway. When B is deficient, the pentose shunt becomes more prominent in degradation of glucose and is often accompanied by excessive synthesis of phenolic acids. Boron also may play a role in translocation of sugars. In plants deficient in B, cell walls are thin and brittle and collapse easily. Patches of cells fail to form in young leaves. Cells elsewhere in the leaf continue to expand, causing the leaf to become crinkled or thickened and leathery. Interference with the formation of cell walls causes many of the other symptoms, such as cracking of stems and notching of leaves. Accumulation of B appears to be more difficult during the reproductive phase than in the vegetative phase of plant development. Thus, when levels of B in the soil are marginal, symptoms may not appear until reproductive development occurs.

An early symptom of B deficiency is reduced growth or death of the apical meristem. Lateral buds produce many small branches (witches'-broom symptom), but their terminal buds die. Primary leaves thicken and become deformed and leathery. Trifoliolate leaves may form only one or two deformed leaflets, and petioles become brittle. Interveinal chlorosis occurs on all leaves. Stems are swollen near nodes. Leaves and stems appear desiccated and may have a stiff, woody feel. Flowers and pods either do not form or they abort; seeds fail to swell, and the root system is poorly developed. Longitudinal cracks appear near the base of the stem. A reddish, corky layer forms in cracked tissue. Symptoms are intensified by low soil moisture.

Boron deficiency occurs rarely in the field but could be corrected by addition of borate salts to soil. However, care should be exercised, since excessive B causes yellowing and necrosis of the margins of primary leaves shortly after emergence and of older leaves. Toxicity symptoms may appear when the soil B content exceeds 5 ppm.

Molybdenum (Mo)

Molybdenum deficiency can occur in acidic soils, especially if Fe and Al contents reduce Mo solubility. The element becomes progressively less available below pH 7.0. Its only known function is as a component of nitrate reductase. Molybdenum deficiency symptoms resemble those of N deficiency. They can be corrected by addition of lime if native Mo availability is being reduced as a result of low pH. Molybdenum deficiency can also be corrected by treating seed with Mo fertilizer.

Toxicity Symptoms

Aluminum (Al)

Aluminum is not required for bean plant growth but may become toxic at high concentrations. Aluminum toxicity is associated with acid soils and is common in the tropics. The solubility of Al increases markedly below pH 5.0, leading to greater concentrations of the element in the soil solution. Aluminum tends to precipitate near roots and restricts the uptake of Fe and Ca. It interferes with phosphate metabolism and causes inorganic phosphate to accumulate in roots while reducing its metabolism and transport. Excess Al may also lead to inhibition of DNA synthesis at root apices, cause conversion of phosphate in the soil to unavailable forms, and inhibit several enzymes involved in respiration and nutrient absorption.

Beans may start to show symptoms of Al toxicity below pH 6.0. Plants affected by excess Al are stunted and have poorly developed root systems with numerous adventitious roots near the soil surface. Older leaves become chlorotic and develop necrotic margins. Plants affected by excess Al may show symptoms of deficiencies in Fe, Ca, and P.

Aluminum toxicity is usually corrected by application of 0.5–2.0 ton/acre (1–5 t/ha) of limestone to the soil. Calcitic lime neutralizes the Al and increases the availability of Ca. Dolomitic limestone provides Mg in addition. Cultivars of dry beans vary in their sensitivity to Al, with some showing moderate tolerance to the element.

Boron

Boron toxicity (Plate 196) generally occurs after nonuniform application of fertilizer or when the fertilizer is band-applied too closely to the seed, especially during dry weather. Boron toxicity may also occur when beans follow a crop, such as turnip, which is normally heavily fertilized with B. Cultivars of dry beans vary in their sensitivity to B deficiency, and some are tolerant.

Manganese

Manganese toxicity occurs in low-pH, volcanic soils. Poor drainage aggravates the problem. Symptoms may appear as purple-black spots on the stem, petiole, midrib,

Table 14.1. Nutrient sufficiency ranges for dry edible beans using upper, fully developed leaf sampled before initial flowering[a]

Nutrient	Sufficiency range
N, %	4.25–5.50
P, %	0.25–0.60
K, %	1.70–3.00
Ca, %	0.35–2.00
Mg, %	0.25–1.00
Fe, ppm	50–450
Zn, ppm	20–70
Mn, ppm	20–100
Cu, ppm	10–30
B, ppm	15–50
Mo, ppm	1–5

[a]Source: Zaumeyer and Thomas (7).

and veins of leaves, especially the lower leaf surface. The pulvinus region is not discolored. Microscopic examinations reveal that these spots are actually clumps of secreted material (MnO_2) around the basal cells of hairs. Chlorosis may develop between major veins, especially on younger leaves. Affected leaves may cup downward and have necrotic margins. Some cultivars are less sensitive to Mn toxicity than others. Improved drainage, addition of organic matter, and lime application may alleviate the problem.

Sodium (Na) and Salinity

Beans do not require Na but are very sensitive to the salinity and/or Na content of the soil. In general, Na content becomes a problem when the percentage of saturation is more than 4% and soil compaction is moderate to severe. Salinity adversely affects beans when the conductivity is more than 0.8 mmho/cm (0.08 S/m). Growth reduction, leaf scorch, and plant death may occur when sensitive bean cultivars are planted in saline soils or soils with a high Na content. Damage may be high during germination and seedling development, and plant stands can be significantly reduced. Correction of salinity problems depends on careful management of the water status of the soil in order to prevent accumulation of salt at or near the soil surface.

Critical and Sufficient Tissue Nutrient Levels

The elemental composition of bean seeds, averaged over three cultivars, has been reported as 3.0% N, 0.54% P, 1.49% K, 0.25% Ca, 0.19% Mg, 0.17% S, 76 ppm Fe, 31 ppm Zn, 19 ppm Mn, 9 ppm Cu, and 12 ppm B. The concentrations of 11 essential elements in normal leaves of dry edible beans are given in Table 14.1. Beans require at least 16 elements for growth. Carbon, hydrogen, and oxygen are supplied by water or air and are not considered further. Nitrogen, P, K, Ca, Mg, and S are required in large amounts and are referred to as macronutrients. Micronutrients required in very small amounts include Fe, Zn, Mn, Cu, B, Cl, and Mo. Nutrient requirements for different cultivars are similar, as judged by responses in fertilizer trials.

The accurate diagnosis of mineral disorders and mineral requirements depends on soil tests and analysis of plant tissues in addition to recognition of visual symptoms on the plant. Leaf analysis can be used as a guide to nutrient deficiencies (Table 14.1). Soil tests by themselves are inadequate as a basis for diagnosis of nutrient deficiencies or toxicities or for fertilizer recommendations. To be used effectively, soil tests and tissue analyses must be correlated with the results of fertility trials. However, soil tests for N characteristically do not correlate with yield, so that recommendations on the N requirements for beans are usually based on factors such as the crop and fertilizer history of the field. Procedures for testing soil and for relating test results to fertilizer requirements for beans have been developed for P, K, Ca, Mg, S, Zn, and Mn. For micronutrients other than Zn and Mn, soil tests are not well developed because deficiencies in beans are not known or are rare in the field. Information on these micronutrients has been obtained largely from nutrient culture studies.

Deficiency symptoms are most commonly seen for the macronutrients N, P, and K and for the micronutrients Zn and Mn. Although fertile soils usually contain adequate amounts of the elements required by beans, N, P, and K are frequently applied at planting, and deficiencies of these elements are common on infertile soils. Beans are generally regarded as having a poor ability to fix N symbiotically, and growth responses to N usually occur in the range of 18–90 lb/acre (20–100 kg/ha). Above a soil pH of 6.8, Ca, Mg, and S levels are likely to be adequate, and deficiencies are rarely reported. Above pH 7.2, deficiencies of Fe, Zn, or Mn may occur, and growth responses to the addition of these elements, or to a reduction of soil pH, may be expected if soil tests for the elements are low. In acid soils, S and P deficiency and Al and Mn toxicity are more likely.

Although nutrient deficiencies and toxicities are usually corrected by application of the appropriate fertilizer or lime, some genetic control is possible, because variability among bean lines in tolerance to Al and Mn toxicity and to P deficiency has been reported.

Selected References

1. Adams, M. W., Coyne, D. P., Davis, J. H. C., Graham, P. H., and Francis, C. A. 1985. Common bean (*Phaseolus vulgaris* L.). Pages 433-476 in: Grain Legume Crops. R. J. Summerfield and E. H. Roberts, eds. Collins, London.
2. Flor, C. A., and Thung, M. T. 1989. Nutritional disorders. Pages 571-604 in: Bean Production Problems in the Tropics. 2nd ed. H. F. Schwartz and M. A. Pastor Corrales, eds. Centro Internacional de Agricultura Tropical, Cali, Colombia.
3. Nuland, D. S., Schwartz, H. F., and Forster, R. L. 1983. Recognition and management of dry bean production problems. Iowa State Univ. North Cent. Reg. Ext. Publ. 198.
4. Scaife, A., and Turner, M. 1983. Diagnosis of Mineral Disorders in Plants. Vol. 2, Vegetables. Her Majesty's Stationery Office, London.
5. Vitosh, M. L., Christenson, D. R., and Knezek, B. D. 1978. Plant nutrient requirements. Pages 94-111 in: Dry Bean Production—Principles and Practices. L. S. Robertson and R. D. Frazier, eds. Mich. State Univ. Ext. Bull. E-1251.
6. Wallace, T. 1961. The Diagnosis of Mineral Deficiencies in Plants by Visual Symptoms. 2nd ed. Chemical Publishing, New York.
7. Zaumeyer, W. J., and Thomas, H. R. 1957. A monographic study of bean diseases and methods for their control. U.S. Dep. Agric. Tech. Bull. 868.

Chapter 15

Albert Ulrich
Department of Soil Science
University of California, Berkeley

Potato

The potato (*Solanum tuberosum* L.) ranks fourth after wheat, rice, and corn as an important source of food worldwide. In the United States, the potato has essentially moved to second place, replacing third-place rice and even second-place corn as a direct food source because of its culinary adaptability and the popularity of french fries at fast food establishments. The potato in its natural state is exceedingly low in sodium and relatively rich in potassium and vitamin C.

As a plant, it thrives in the temperate zone in many areas differing greatly in climate and in soil composition but still produces large tonnages of tubers, which can be shipped directly to market or stored until needed for distribution or processing.

To produce large yields, the plant requires a steady stream of nitrogen (N), along with adequate supplies of phosphorus (P), potassium (K), and the remaining nutrients, calcium (Ca), magnesium (Mg), sulfur (S), iron (Fe), zinc (Zn), manganese (Mn), copper (Cu), boron (B), and molybdenum (Mo). Ordinarily, these nutrients are provided by the soil during the 90- to 120-day growth period of the crop, but if supplies are not adequate as shown by stress symptoms or by plant analysis results, nutrients can be added as fertilizer at appropriate times from inorganic and/or organic sources. They can be provided indirectly by changes in cultural practices, such as by introducing irrigation or changing irrigation practices, planting cover crops, rotating crops, removing weeds, reducing the number of plants per unit area, and planting improved varieties.

Five Important Nutritional Steps

The detection, correction, and prevention of plant nutrient deficiencies of potatoes can be described in five important steps: 1) identify the symptoms visually by using the descriptive key and color plates that accompany chapter; 2) confirm the visual diagnosis by comparing the analytical results of blades or petioles from comparable leaves, with and without symptoms, in Table 15.1; 3) conduct small-scale field trials using appropriate materials as a corrective measure; 4) compare the analytical results of blades or petioles taken from the field trials at 2-week intervals after fertilization and irrigation or after significant rainfall to verify that the newly applied fertilizer reached the leaves for new growth; and 5) follow a systematic plant analysis program, thus preventing nutrient deficiencies and yield losses of current and subsequent crops. Only by local experience can the potato crop now so widely grown on so many soil types be fertilized effectively for best crop production. The color plates, table, and figures have been presented in this chapter to enable the potato grower to meet this goal.

Visual Diagnosis

Signs of Stress

Signs of stress will appear whenever any one of the mineral nutrients becomes severely deficient beyond the "marginal deficiency stage" of development. Each element develops its own characteristic stress symptoms, such as overall yellowing for N, S, and Mo; leaf scorch of young mature leaves for K and Mg; growing-point damage for Ca and B; stunted greening for P; and initial yellowing of young immature leaves followed by green veining for Fe, dark to black spotting for Mn, fern leaf for Zn, and leaf roll for Cu deficiency. These symptoms are presented in the color plates.

Healthy Plant

A normal healthy potato plant can be readily recognized during the grand growth period by an experienced potato grower (Plates 197–199). It is a plant with a normal green color, with a good sturdy top-growth. It tests high in nutrients, particularly for nitrate-N in the petioles, as can be shown in the field by the diphenylamine reagent (see Plate 232 and the Appendix to this chapter) or by the more accurate laboratory methods.

Abnormal Plants

A potato grower becomes concerned about a potato crop whenever symptoms appear that cannot be diagnosed as caused by disease, insects, or mechanical damage, or by other causal factors such as low air and soil temperatures, water shortage, poor aeration caused by soil compaction, poor drainage, soil salinity, soil acidity, frost, residual herbicides and other pesticides,

etc. At times, such factors can decrease nutrient supplies to the point that deficiencies occur on an otherwise fertile soil. The remedy is to correct the primary cause of the deficiency and then have another look at the nutrient balance within the plant by using plant analyses. Fertilizing with the required nutrients will overcome the nutrient deficiency and achieve better crop production. Thereafter, the potato crop should be monitored for

Table 15.1. Plant analysis values for determining the mineral status of potato (dry basis)

Nutrient	Plant part tested	Tentative critical concentration[a,b]	Range showing symptoms Deficiency[c]	No deficiency[d]
Boron, ppm	Blade	20[e]	1.2–20	>20
Calcium, %	Petiole	0.15	0.09–0.15	0.15–2.5
	Blade	0.15	0.05–0.15	0.15–2.5
Chlorine, %	Petiole	...	...	...
Copper, ppm	Blade	5[e]	<5	>5
Iron, ppm	Blade	35[e]	15–35	50–400
Magnesium, %	Petiole	0.06	0.04–0.06	0.06–1.00
	Blade	0.09	0.04–0.09	0.10–1.00
Manganese, ppm	Blade	25	4.0–25	40–350
Molybdenum, ppm	Blade	15	...	...
Nitrogen (N) NO_3^--N, ppm	Petiole	2,000	0–500	1,000–38,000
	Blade	300	0–400	500–11,700
Total N, %	Petiole	1.6	1.1–1.5	1.6–4.25
	Blade	2.5	1.9–2.9	3.0–6.00
Phosphorus (P) $H_2PO_4^-$-P, ppm	Petiole	1,000	400–750	1,000–8,000
	Blade	1,200	500–1,000	1,400–4,500
Total P, ppm	Petiole	1,200	600–1,200	1,400–11,000
	Blade	1,800	1,000–1,750	2,500–12,500
Potassium, %	Petiole[f]	2.3	0.3–1.8	3.0–11.5
	Blade	1.0	0.3–0.9	1.5–5.0
Sodium	Petiole[f]	...	...	0.04–0.44
	Blade	...	...	0.01–0.16
Sulfur (S) SO_4^{2-}-S, ppm	Blade	250	30–250	250–1,000
Total S, ppm	Blade	750	400–750	800–3,000
Zinc, ppm	Blade	20[e]	8–20	20–150

[a] The critical concentration of a nutrient for plants grown hydroponically is usually taken at the point in a calibration curve where growth is 5 or 10% less than the maximum.

[b] All critical concentrations listed are based on a "young mature leaf" sample usually designated as the second leaf below the flat top (Fig. 15.2) or as the fourth leaf below the "youngest leaf."

[c] Leaf material for chemical analysis must be collected shortly after appearance of leaf symptoms, otherwise deficient plants may accumulate nutrients in the leaf without restoring chlorotic tissues to normal. Petioles of old yellow leaves from N-deficient plants often give a high nitrate test, even though petioles from younger light green and green leaves on the same plant give a very low nitrate test.

[d] The upper value reported is the highest value observed to date for normal plants. Abnormally high values are often associated with other nutrient deficiencies. For example, blades low in Fe may contain up to 4% Ca and/or 900 ppm of Mn. Also, blades deficient in Zn may contain as much as 850 ppm of Fe.

[e] Critical values based on leaves with mild deficiency symptoms developed hydroponically under clear glass, smogfree greenhouse conditions.

[f] For best results, petioles should be analyzed for K, even though petioles under low-K conditions may retain K concentrations up to 2.0% K when corresponding blades show K deficiency symptoms at 0.6% K.

nutrients through a systematic plant analysis program. This is an effective way of preventing marginal stress and extreme nutrient deficiencies and maintaining an adequate nutrient supply for growth from emergence to harvest. If, for example, a shortage of N is detected early in the growing season on a light, sandy soil low in organic matter, fertilizing with N can frequently enhance yields greatly. Similar responses can also be obtained with other nutrients with only a slight increase in variable costs. Well-nourished plants when fertilized according to their needs based on plant analysis will lead to an excellent crop of potatoes with only a minimum or no movement of fertilizer to ground water, a point of concern in some localities.

Diagnostic Procedure

A visual diagnosis of symptoms should be made as soon as possible after the symptoms have appeared. Later on, the symptoms cannot be easily classified because they will have been modified by other factors. Comparable leaf samples, with and without symptoms, should be collected for chemical analysis when symptoms first appear, because the plants may outgrow the symptoms. For example, higher soil temperatures may obliterate a low-temperature-induced P deficiency. When look-alike symptoms have appeared, such as leaf scorch caused by drought or sunburn, they can be separated by chemical analysis from those of K and Mg deficiency or even to separate a K deficiency from a Mg deficiency. When the symptoms have been identified correctly, the leaf levels will be as low as the values given in Table 15.1.

Deficiency Symptoms

Uniform Yellowing

Nitrogen

The entire potato plant turns a light green to pale yellow when it first becomes N deficient (Plates 200–202). As the deficiency increases in severity, the leaves at the apex become smaller and curl upward. At these times, the petioles give a negative test for nitrate-N with diphenylamine reagent (Plate 232). The lower, older leaves become deep yellow to light brown to necrotic, and when dry they separate from the stem easily. Trapped nitrate in old petioles will test positive for nitrate. In the meantime, the apical leaves may turn green because of nitrification of soil organic matter, followed by nitrate absorption and conversion of nitrate into amino acids and amides for growth at the apex. Petioles of these leaves will often test negative for nitrate. In this scenario, top growth is retarded more than root growth so that the top-to-root ratio declines substantially along with a major decrease in tuber size. A side-dressing of ammonium nitrate or some other readily available N source will correct the N deficiency. Nitrogen deficiency

is indicated under field conditions when petioles from recently matured leaves as shown in Plates 201, 202, and 205 contain less than 500 ppm of nitrate-N (dry weight basis).

Sulfur

Sulfur deficiency is similar to N deficiency except that the entire plant, including the younger leaves, remains light green with time (Plates 203 and 204). Severely deficient leaf blades remain yellowish and in time curl upward. Petioles of S-deficient plants give a strong positive test for nitrate with diphenylamine reagent, turning a deep blue color instantly instead of remaining colorless as in comparable petioles from N-deficient plants (Plate 232). Sulfur deficiency is indicated when leaf blades contain less than 250 ppm of sulfate-S or less than 750 ppm total S (dry weight basis).

Molybdenum

The uniform yellowing of leaf blades in Mo deficiency is similar to N and S deficiencies (Plate 205). A diphenylamine test for nitrate with Mo deficiency will be strongly positive (Plate 232). Based on sugar beet values, Mo deficiency could be expected in potatoes when leaf blades contain less than 0.15 ppm of Mo (dry weight basis).

Stunted Greening

Phosphorus

Phosphorus deficiency symptoms of potato are not readily recognizable (Plates 206–208). The plants appear somewhat stunted and have a darker green color than adjacent nondeficient plants. As P deficiency increases in severity, leaf roll, the upward cupping of the leaf blades, occurs and reveals the gray-green color of the lower surface. Phosphorus deficiency occurs frequently when soil temperatures are relatively low, 50°F (10°C) or lower. Potatoes, like tomatoes and sugar beets, are likely to respond to preplant P applications or to a delay in planting until the soil temperatures rise above 60°F (15°C). Phosphorus deficiency frequently appears on acid and calcareous soils or on kaolinitic soils. Since the symptoms are not readily recognizable, shortly after emergence plant samples should be taken at 2-week intervals and continued to harvest. Phosphorus deficiency takes place when petioles or blades contain less than 750 ppm of H_2PO_4-P soluble in 2% acetic acid or less than 1,750 ppm of total P (dry weight basis).

Leaf Scorch

Potassium

Diagnostically it is important to note that the potato is a sodium (Na) excluder, and this characteristic results in a relatively high K concentration in the petiole even when the blade is visibly deficient in K. At this time, the petiole will give a false positive K value, i.e., be high in K concentration even when the blade is clearly deficient in K (Table 15.1). As an Na excluder, K deficiency symptoms appear first on young, fully expanded leaf blades as a glossy sheen with pronounced crinkling and slightly black pigmentation (Plates 209–211). A severe deficiency causes marginal leaf scorch, which gradually expands as dead tissue and final loss of leaves. Youngest leaves near the growing point remain green and relatively high in K content. Potassium deficiency may occur in plants on sandy loam soils with low exchange capacity; on clay loam soils with a high unavailable K fixing power; on serpentine soils high in exchangeable Mg; or on peat, muck, and other soils releasing K too slowly during the grand growth period of the crop. However, micaceous soils with low replaceable K, but high in Neubauer K, fail to respond to K even when fertilized with ample N and P. Potassium deficiency is indicated whenever petioles contain less than 2.0% K or blades contain less than 1.0% K (dry weight basis). Leaves with more than 1.0% Na indicate root damage or dust on leaves or even contamination from hands.

Magnesium

Magnesium deficiency symptoms appear first on young mature leaves as a slight chlorosis with green veining and brown spotting, which terminate as interveinal leaf scorch, in contrast to the marginal scorch of K deficiency (Plates 212–214). Symptoms are most severe in the oldest leaves, which have green veins and large patches of marginal and interveinal necrosis. Leaflets of severely stressed leaves become dry and turn brown, but most remain attached to the rachis. In both deficiencies the leaves near the growing point remain green. On some soils, Mg deficiency symptoms may be induced by the frequent use of K without including Mg in the fertilizer program. Magnesium deficiency is indicated when petioles or blades from leaves with symptoms contain less than 0.06 or 0.09% Mg, respectively (dry weight basis).

Growing-Point Damage

Calcium

Calcium deficiency symptoms appear first at growing points of the potato plant (Plates 215–217). When the plants are grown hydroponically, the root tips of Ca-deficient plants become deformed and dark in color. These symptoms occur even when the Ca concentration in the nutrient solution still ranges from 3 to 8 ppm. Shortly thereafter, tipburn appears on the youngest center leaves including the growing point of the plant. Severe deficiencies cause young mature leaflets to roll upward with chlorosis and brown spotting. Since Ca movement within the potato plant is primarily unidirectional, a continuous supply of Ca is required to prevent

Ca deficiency of growing points and for tuber formation. These needs for Ca can be met by appropriate additions of limestone to acid soils or of gypsum to sandy soils. Calcium deficiency of potatoes is indicated when petioles or blades contain less than 0.15% Ca (dry weight basis).

Boron

Boron deficiency of potato causes the formation of a bushy plant with droopy leaves (Plates 218–220). Blades crinkle, cup upward, and are bordered by light brown tissue. Boron deficiency, like Ca deficiency, affects the growing points. Root tips become swollen and darken; immature center leaves become deformed, and the growing point dies. Unless local experience indicates otherwise, try small-scale field trials with sodium borate (ordinary borax) applied at the rate of 10–15 lb/acre (11–17 kg/ha) preplant or later as a side-dressing when early symptoms have appeared. Boron deficiency is indicated when deformed blades contain less than 20 ppm of B (dry weight basis).

Yellowing with Green Veining

Iron

Iron deficiency of the potato grown hydroponically appears initially as a yellowing of young leaves near the growing point of the plant (Plates 221–223). In time, the leaves become light yellow to nearly white, which is followed occasionally by a blotchy, light brown necrosis. During the deficiency stage, blade tips remain green the longest. Netted green veining occurs when traces of Fe are absorbed and translocated along the veins for chlorophyll formation. Green veining is actually an Fe recovery symptom. On acid soils, Fe deficiency is unlikely to occur, but when symptoms appear on calcareous soils, foliage sprays of chelated Fe solutions should overcome the deficiency. Iron deficiency is indicated when blades contain less than 35 ppm of Fe (dry weight basis).

Zinc

Zinc deficiency of potatoes, often known as "fern leaf" or "little leaf," appears on young developing leaves as a chlorosis, followed by the formation of narrow, upwardly cupped leaf blades, often with tipburn (Plates 224–226). As the plants grow, blotching and necrosis increase, and the petioles (leaf stalks) detach easily from the stem, giving the plants a profile similar to a palm tree. Zinc deficiency is less likely to appear on acid soils than on calcareous soils. Without local experience in correcting Zn deficiency of potatoes, try on a small-scale foliage sprays containing Zn or soil applications of Zn sulfate or Zn chelates as soon as symptoms have appeared. Zinc deficiency is indicated when blades contain from 8 to 20 ppm of Zn (dry weight basis).

Yellowing of Young Leaves

Manganese

The first sign of Mn deficiency is a yellowing and slight cupping of younger leaves (Plates 227–229). With increased deficiency, dark to black spotting develops between the veins with a large increase in spotting appearing along larger veins and the midrib. These symptoms, darkening and cupping, increase in severity with time.

Manganese deficiency is most likely to occur on calcareous soils and least likely on acid soils. Manganese toxicity is most likely to occur on acid soils and least on calcareous soils. As a corrective measure on calcareous soils, spray the foliage in a small-scale trial with a recommended commercial solution to overcome Mn deficiency. On acid soils, where toxicity occurs, raise the soil pH by liming the soil. Manganese deficiency is indicated when blades contain less than 25 ppm of Mn (dry weight basis).

Copper

An early sign of Cu deficiency is the development of a uniform, light green color of young, immature leaf blades (Plates 230 and 231) similar to those of S, Mo, Mn, and Fe deficiencies. Thereafter, Cu deficiency is manifested primarily by a pronounced upward cupping and inward rolling of the young, relatively large, leaf blades. This is in sharp contrast to the small, narrow leaf blades of Zn deficiency. Copper deficiency is indicated when blades with symptoms contain less than 5 ppm of Cu (dry weight basis).

General Considerations

When the visual diagnosis has been completed, the nutrient judged to be deficient should be confirmed by chemical analyses of leaves with and without symptoms as given in Table 15.1.

Based on local experience, a nutrient deficiency may be corrected. It is usually done on a small-scale, test-plot basis by fertilizing the current or succeeding crops judiciously on the same or adjoining fields. On a test-plot basis, appropriate leaf samples should be taken before fertilization and sampled again 2 weeks after irrigation or after ample rainfall. Sampling should be continued thereafter at 2-week intervals to harvest. The mere addition of fertilizer is no proof that nutrients become available for better plant growth.

How can the nutrient requirements of the potato crop be met on a sustained yield basis, especially when soils within the temperate zone vary so widely in pH, texture, depth, mineral composition, and nitrification of organic matter? One way is to take out fertilizer insurance by buying a complete fertilizer and using it liberally or by applying generous amounts of barnyard manure to supply mainly N and to improve soil texture. This program often leads to the excessive use of fertilizer and possible ground water pollution. Another extreme way is to use no fertilizer and accept the yield as produced from the natural fertility of the soil. Unfortunately, this program often leads to extreme N deficiency. Nitrogen is often the key element in a fertilizer management program. When the demand for potatoes as a food source increased dramatically, field trials demonstrated almost immediately that N increased yields greatly, and if a little N was good, more was believed to be better. How much N is actually needed by the crop as fertilizer, not supplied by the soil, is the question to be answered, not only for N but also for P, K, Zn, and the other nutrients, for best crop production on a sustained yield basis?

Thus, in a five-step diagnostic program, ask the plant how it is doing nutritionally by 1) using visual symptoms; 2) using chemical tests such as for nitrate with diphenylamine reagent (Plate 232) or the more accurate laboratory methods (Table 15.1); 3) using field trials that measure yield responses; 4) comparing visual symptoms, if any, and analytical results (Table 15.1) of leaf samples taken from the field trials; and 5) monitoring each production unit by plant analysis every 2 weeks from emergence to harvest (Fig. 15.1). Thus, in conventional or nonconventional agriculture, a favorable fertilizer program can be adjusted for the current crop when time permits or for succeeding crops on the same field.

Changes in a fertilizer program also require information about the soil, climate, and pest and disease factors, as well as about the fertilizer materials to be applied, and methods, timing, and frequency of

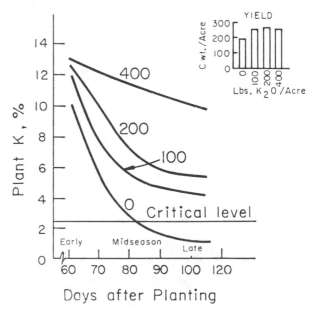

Fig. 15.1. Petiole K levels and tuber yields of cv. White Rose potatoes fertilized with four rates of K. Redrawn from field experiments from Tyler et al (7).

application. These are all important in meeting the nutrient needs of the crop successfully. The effects of these and other cultural changes on the mineral nutrition of the potato are answered for the potato grower in a systematic plant analysis program on a sustained yield basis.

Plant Sampling and Analysis

After emergence, select a young mature leaf just below the "flat top" for chemical analysis (Fig. 15.2), and thereafter collect samples at 2-week intervals until harvest. This leaf will be of the same physiological age for all samplings, and therefore will have comparable critical values wherever and however the crop is grown (Table 15.1).

The second leaf below the crown or flat top of the plant shown in Figure 15.2 is the leaf to sample to determine the nutrient status of the potato plant. This leaf is often referred to as a "young mature leaf" or as a "recently matured leaf." The terminal "bladelet" of this leaf is taken as the "blade," and the "petiole-rachis" without bladelets is designated as the petiole. Take a single leaf per plant at 20–40 equidistant intervals across the rows of one-quarter of a 10- to 40-acre (4-

to 16-ha) field. The results of four such samples, one per quarter, will indicate the overall variability of the field and of the analytical determinations. Plotting the analytical results for each sampling date against time relative to the critical concentration will reveal how well the fertilizer program has met the nutrient needs of the potato crop (Fig. 15.1). With this information plus local experience, a grower can adjust his fertilizer program to meet the needs of the crop effectively.

The labor-intensive analytical methods described in Johnson and Ulrich (4) involving wet chemistry and low-cost equipment have been superseded by expensive but precise and rapid methods such as atomic absorption spectroscopy or inductively coupled plasma emission spectroscopy for cations K, Ca, Mg, Fe, Mn, Zn, Cu, and Na, plus total P, total S, and total B. The anions NO_3-N, SO_4-S, H_2PO_4-P, and Cl can be determined by ion chromatography or by specific ion electrodes, and the Kjeldahl total N minus NO_3-N by an ammonium and nitrate autoanalyzer. A reference sample should be run in triplicate with each set of analyses to detect possible analytical errors before they become serious.

Critical and Sufficient Tissue Nutrient Levels

The critical nutrient concentration is a convenient reference point for assessing the nutrient status of a crop, providing a "recently matured" leaf is taken as the sample for analysis. A recently matured leaf can be taken any time during the growing season, since it is approximately at the same physiological age not only for a single

Fig. 15.2. Potato leaf sample identification. The immature leaf is the oldest leaf on the flat top. The second leaf is referred to as the recently matured leaf. Reprinted, by permission, from Fong and Ulrich (1).

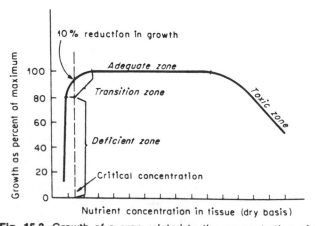

Fig. 15.3. Growth of a crop related to the concentration of a nutrient in the tissue. The critical concentration is usually taken at the point where growth is 10% less than the maximum. Deficiency symptoms usually appear below the critical concentration and not at all above it. The sharper the transition zone, and the broader the range of values from deficiency to toxicity, the more useful the calibration curve becomes for diagnostic purposes. Toxicity symptoms appear in the toxic zone as a nutrient becomes excessive, for example, when potato petiole nitrate-N exceeds 34,000 ppm (dry basis). Reprinted, by permission, from Ulrich and Hills (9).

sampling but for all samplings during the growing season.

The critical concentration (Fig. 15.3) can be conveniently determined hydroponically where all nutrients except the one to be calibrated are adequately supplied for growth. The calibrated nutrient is supplied one time only in convenient doses, starting from nil to a full treatment such as in half-strength modified Hoagland's nutrient solution. The plants are harvested when those in the first half of the series are visibly deficient and the remainder are without symptoms.

The critical concentration is taken at the point where growth is 5 or 10% below the maximum for plants that are not deficient. Because of the greater variability of field sampling, the practical critical concentration is usually set at a higher value than that determined under controlled greenhouse conditions.

The safe level is the nutrient concentration maintained appreciably above the critical concentration during the growing season up to harvest for best crop production, with the possible exception of N. Here, a deficiency of N several weeks before harvest, as with sugar beets, frequently improves the quality of the crop for processing or storage for later use.

Unfortunately, by the time nutrient stress symptoms appear, especially when widespread, significant crop losses have already occurred. Crop monitoring by a systematic plant analysis program can prevent such crop losses, as well as detect marginal deficiencies, before visual symptoms appear.

Management Considerations

Most potato growers have a fertilizer program based on local experience. Their program can be evaluated by yields alone or by a combination of yields and plant analysis results. Both procedures have predictive powers. With yield results alone, the predictions are restricted to the original location or to local fields with similar soils and culture practices. By comparing plant analysis and yield results similar to those shown for K (Fig. 15.1), field predictions can be made wherever potatoes are grown. (Plants are sampled at 2-week intervals, and the results are compared with an appropriate critical level, e.g., 2.3% K in petioles of a young mature leaf, which is defined as the second leaf below the "flat top" of the potato plant as shown in Fig. 15.2. This is also the fourth leaf down as recommended by Tyler et al [7]).

Evaluating a fertilizer program for the potato is a wait-and-see situation without making dramatic changes in midstream. For example, the initial petiole K values of 10–13% shown in Figure 15.1 were dramatically above the critical value of 2.3%, and hence no response to K was anticipated. This hypothesis had to be adjusted, however, as subsequent results for the untreated plots decreased from 10% K to the critical value of 2.3% K,

with the next two treatments not too far behind. On the next sampling, the K supply of treatments above 100 lb/acre of K_2O (110 kg/ha) surprisingly met the demand by remaining well above 2.3% K, and therefore no yield increase above 100–150 lb/acre of K_2O (110–170 kg/ha) was anticipated or observed. On the basis of the yield and plant analysis results (Fig. 15.1), each grower on a similar soil with a similar cropping history could expect 100–150 lb/acre of K_2O (110–170 kg/ha) to meet the K needs of the next potato crop. Whatever fertilizer adjustment the grower makes, each field should continue to be sampled on a yearly basis. First, evaluate the current season's fertilizer program, and second, continue or adjust the fertilizer program for the next potato crop.

A systematic plant analysis program can not only prevent nutrient deficiencies but also avoid overfertilization. Adding fertilizer as insurance when the soil nutrient supply is already adequate is uneconomical and often politically unacceptable when it is perceived as a possible pollution source.

For best crop production, there is no easily recognizable nutrient balance within the plant except when an essential nutrient becomes so low it limits growth. When N is limiting, there is no substitute for N. This can be easily shown by growing plants hydroponically in a closed system with a single dose of N when other nutrients are in ample supply. At these times, nutrient ratios vary widely without affecting yield. Rapid growth is restored only when N is replenished.

Predicting the outcome of a fertilizer program based on a single plant sample taken early in the growing season is difficult primarily because of the unpredictability of climate and its effects on on subsequent plant growth. In a favorable climate thereafter, plant nutrient demand often may exceed soil nutrient supply. This could induce plant nutrient deficiency and possible crop loss. Conversely, in an unfavorable climate, the soil nutrient supply may exceed demand and, in the case of N, excesses may lead to a lower quality potato. In a new production area, it is initially better to sample fields more often than only once during the growing season, particularly for the rapidly growing, shallow-rooted potato plant. Also, the longer the time interval between sampling and harvest, the greater the likelihood of estimating the fertilizer needs of the crop incorrectly.

Selected References

1. Fong, K. H., and Ulrich, A. 1969. Leaf analysis as a quick guide for potassium nutrition of potato. J. Am. Soc. Hortic. Sci. 94:341-344.
2. Harris, P. M., ed. 1982. The Potato Crop. Chapman and Hall, New York.
3. Hooker, W. J., ed. 1981. Compendium of Potato Diseases. American Phytopathological Society, St. Paul, MN.
4. Johnson, C. M., and Ulrich, A. 1959. Analytical methods

for use in plant analysis. Pages 25-78 in: Calif. Agric. Exp. Stn. Bull. 766.

5. Ng, E., and Loomis, R. S. 1984. Simulation of growth and yield of the potato crop. Pudoc Simulation Monographs, Wageningen, The Netherlands.

6. Rhoades, R. E. 1982. The incredible potato. Natl. Geogr. 161:668-694.

7. Tyler, K. B., Lorenz, O. A., and Fullmer, F. S. 1961. Plant and soil analysis as guides in potato nutrition. Calif. Agric. Exp. Stn. Bull. 781.

8. Ulrich, A., and Fong, K. H. 1969. Effects of potassium nutrition on growth and cation content of potato leaves and tubers relative to plant analysis. J. Am. Soc. Hortic.

Sci. 94:356-359.

9. Ulrich, A., and Hills, F. J. 1990. Plant analysis as an aid in fertilizing sugarbeet. In: Soil Testing and Plant Analysis. 3rd ed. R. L. Westerman, ed. Soil Sci. Soc. Am. Book Ser. 3.

10. Ulrich, A., Mostafa, M. A. E., and Allen, W. W. 1980. Strawberry deficiency symptoms: A visual and plant analysis guide to fertilization. Univ. Calif. Publ. Bull. 1917.

11. University of California Statewide Integrated Pest Management Project. 1986. Integrated Pest Management for Potatoes in the Western United States. Publ. 3316. University of California Division of Agriculture and Natural Resources, Oakland.

Appendix
Notes on the Preparation and Use of Diphenylamine

Diphenylamine test for nitrate. Place a drop of diphenylamine reagent on a slantwise cut surface through the petiole of a recently matured leaf (Plate 232). If a distinct blue color develops immediately, the nitrate-N value is more than 500 ppm (dry basis), and the plant is adequately supplied with N at the time of the test. If the reagent remains colorless, turns blue very slowly, or becomes brown rapidly, the nitrate-N value is less than 500 ppm, and the plant is therefore N deficient at the time of sampling. See caution below on the use of the diphenylamine reagent.

Preparation of diphenylamine reagent. Add 0.2 g of diphenylamine to 100 ml of nitrate-free, concentrated sulfuric acid. Store reagent in a fully labeled Pyrex glass-stoppered bottle and keep in the dark. Dispense small amounts of the reagent as needed from a dark-colored glass-stoppered eyedropper bottle for field use. Add a few crystals of table salt (NaCl) to freshly prepared reagent if it fails to develop color in the presence of nitrate.

Caution: Sulfuric acid is highly caustic. Keep away from children and animals. Do not spill on clothing. If reagent contacts skin, wash immediately with large amounts of water. Follow this with a dilute solution of baking soda. *Always pour unused acid or reagent into water, not the reverse.*

Part IV

Fruit Crops

Chapter 16

Eric Hanson
Horticulture Department
Michigan State University, East Lansing

Apples and Pears

The nutrient requirements of apple (*Malus domestica* Borkh.) and pear (*Pyrus communis* L.) have been studied extensively throughout the 20th century. Descriptions of nutrient deficiency symptoms have been published for pear (12), apple (11), and tree fruit crops in general (33). Photographs of nutrient-related symptoms on apple and pear have also been published (5,17,40).

Apple and pear trees are fertilized to maintain adequate vigor and to optimize fruit yield and quality. Nitrogen applications usually have the greatest effect on these characteristics and are required in most orchards annually. Applications of other nutrient elements on apples and pears are generally made only when soil or tissue analyses or tree appearance indicates a need.

Apple and pear trees are deep-rooted relative to most crop plants and are able to obtain adequate amounts of most nutrients, even from soils that test low in nutrients. For example, trees often perform well on soils containing levels of P that are inadequate for optimum growth of many annual crop species.

The yearly nutrient requirement of apple and pear trees is difficult to estimate. The quantity of nutrients utilized by annual crops can be measured by analyzing whole mature plants. Perennial crops store nutrients from one year to the next, so the annual demand of fruit trees is often only estimated by measuring the quantities of various nutrients in fruit and other plant parts. Some estimates of the amounts of nutrients removed from mature orchards in the harvested fruit are summarized in Table 16.1. Although these estimates are influenced by factors such as crop load, tree age, and plant spacing, they may provide a useful measure of the quantity of nutrients removed from orchards annually.

The quantities of nutrients required annually for vegetative growth are more difficult to estimate. Leaves in a mature 1-acre apple orchard may contain 42 lb (19 kg) of N, 3 lb (1.3 kg) of P, 47 lb (21.3 kg) of K, 76 lb (34.5 kg) of Ca, and 16 lb (7.2 kg) of Mg (2). One-year-old twigs in the same 1-acre orchard (sampled in the autumn) may accumulate 1.1 lb (0.5 kg) of N, 0.2 lb (0.1 kg) of P, 0.5 lb (0.2 kg) of K, 2.2 lb (1.0 kg) of Ca, and 0.4 lb (0.2 kg) of Mg (38). Considering these values with levels in harvested fruit (Table 16.1) provides a reasonable estimate of the annual nutrient demands of mature orchards. One component that is difficult to estimate is the quantity of nutrients retained in the lateral growth of branches and trunks as well as roots.

Deficiency Symptoms

Nitrogen (N)

Apple and pear trees deficient in N exhibit reduced terminal growth and leaf size (Plate 233). Leaf color is pale green (apple) to bronze (pear) and similar in appearance to normal leaves senescing in the autumn. Color is uniform across the leaf, with no mottling. Older leaves at the base of shoots and spurs develop chlorosis earliest and may abscise prematurely. New twigs are usually thin, with light brown to reddish bark.

Fruit on N-deficient trees are smaller and may color and mature earlier than normal. Fruit set may be poor and fruit drop may be excessive.

Phosphorus (P)

Although P deficiency is uncommon in apple and pear orchards, symptoms of deficiency have been induced on young trees under greenhouse conditions (6,8). Deficient trees exhibit reduced shoot growth and flowering. Bud break may be delayed and fruit set will be poor. Fruit mature early and are often misshapen, russeted, or cracked.

Leaves on P-deficient trees develop a characteristic dark green to purple color, which is usually most intense in or near the main veins of the leaf (Plate 234). Trees slightly deficient in P may not show purple tints in leaves until late summer or autumn. If the deficiency is severe, symptoms may develop soon after bud break. Leaves are smaller than normal and remain somewhat upright, with a narrow angle between the petiole and shoot. As the deficiency progresses, leaves turn light green or yellow, and older leaves abscise early (Plate 235).

Apparent symptoms of P deficiency were observed in pear orchards in Colorado (14). These included a scorching of leaf margins, decreased terminal growth and leaf size, knotty misshapen fruit, scaly bark, and dieback of shoot terminals. Raese (29) was able to increase growth and yield in Washington apple orchards by treating with P fertilizers. Untreated trees exhibited decreased shoot growth and leaf size. Leaf curling and reddish bark were also noted.

Potassium (K)

Potassium deficiencies are common in many apple and pear production regions. The symptoms of K deficiency, first described on apple trees in pot cultures by Wallace (39), were later observed and corrected by K applications on trees in the field (15).

Potassium deficiency is characterized by a scorching of leaf margins. Margins may first appear light green and later turn necrotic. Necrosis starts at the margin and progresses inward toward the midrib. Leaves may appear tattered as necrotic areas fall off. Symptoms may also be accompanied by leaf abscission. Scorching is typically most severe on older basal leaves and becomes less apparent on the most apical leaves. Symptoms usually develop late in the season on slightly deficient trees or earlier if the shortage is acute. In some instances, K deficiency may reduce the color (16) (Plate 236) and acidity (20) of apple fruit.

Although K toxicity has not been reported in apple or pear, excessive levels of K may reduce Mg absorption and induce a deficiency of this element (9).

Calcium (Ca)

Vegetative symptoms of Ca deficiency have been induced by removing the supply of Ca from young apple trees in pot culture (6). Trees rapidly cease shoot and root growth. Shoot tips also exhibit dieback, and viable roots thicken abnormally. Leaf symptoms develop first near the terminal leaf margins but may extend into interveinal tissue. Affected areas turn chlorotic and eventually necrotic. A purplish tint may also develop, appearing first on minor veins and eventually on larger veins.

These foliar symptoms are seldom seen on trees in the field. However, inadequate Ca in the fruit of apple and pear is associated with various disorders and reduced fruit quality. For this reason, the Ca nutrition of apple and pear is of great importance and has been studied extensively. Fruit disorders associated with inadequate Ca include bitter pit (Plate 237), cork spot, internal breakdown, senescent breakdown, and water core in apple and cork spot of pear (19,32).

Magnesium (Mg)

Descriptions of Mg deficiency symptoms have been published for apple (10,27,35) and pear (23,28) and are generally similar for both crops (Plates 238 and 239). Symptoms appear first as pale regions between the main veins of older shoot or spur leaves. Faded areas normally turn brown (apple) to dark brown (pear) and eventually die, leaving necrotic regions. The necrotic areas usually progress towards the leaf margins in apples or may be confined between the veins towards the interior of pear leaves. Symptoms may not appear uniformly on all branches.

Leaves abscise prematurely on severely affected trees, often leaving only a few leaves at the tip of shoots later in the season. Acutely deficient trees show symptoms early in the season and may be entirely defoliated by September.

Fruit on affected trees are often small and poor in quality. Early fruit drop is common. The effects of low Mg on fruit appear to be caused by the reduction in leaf surface and photosynthetic capacity.

Sulfur (S)

Sulfur deficiencies are extremely uncommon in apple and pear production, although some commercial apple orchards in the Pacific Northwest have responded to S applications (4). Orchards exhibiting S deficiency symptoms were on soils of variable pH and texture.

Symptoms of S deficiency include a uniform chlorosis of leaves, which appears similar to N deficiency (Plate 240). Unlike N, symptoms of S deficiency occur first on younger leaves. Yellow leaves eventually develop necrotic areas near the margin. Tree growth is reduced if symptoms are severe.

Iron (Fe)

Iron deficiency occurs on calcareous soils, predom-

Table 16.1. Quantities of major nutrient elements removed annually in apple and pear fruit

Cultivar	Nitrogen		Phosphorus		Potassium		Calcium		Magnesium		Reference
	lb/acre	kg/ha	lb/acre	kg/ha	lb/acre	kg/ha	lb/acre	kg/ha	lb/acre	kg/ha	
Apple											
Delicious	18.4	20.6	5.7	6.4	50.4	56.5	3.9	4.4	2.0	2.2	2
Golden Delicious	18.9	21.2	3.6	4.0	106.8	119.6	3.9	4.4	3.3	3.7	24
Baldwin, Greening	19.0	21.3	3.7	4.1	37.2	41.7	2.8	3.1	3.8	4.3	38
Cox's Orange Pippin	9.3	10.4	2.8	3.1	30.0	33.6	1.1	1.2	1.2	1.3	22
Pear											
Kieffer, Duchess	9.0	10.1	1.4	1.6	16.7	18.7	1.6	1.8	1.6	1.8	38

inantly in arid regions, and is characterized by a distinct pattern of chlorosis, developing first on the actively growing leaves on shoot tips (Plate 241). Major and minor veins remain green while interveinal tissues turn chlorotic to near white. Necrotic regions may develop along the margins or interior of leaves. Severely affected leaves abscise prematurely, and shoot extension and diameter are reduced.

Zinc (Zn)

Deficiencies of Zn have been observed in apple and pear orchards regularly. The most common symptoms are termed "little leaf" or "rosette" (Plates 242 and 243). Small, distinctly narrow leaves are produced at the shoot terminal. Shoot elongation is reduced so that a tuft or rosette of leaves develops at the terminal. Leaves behind the rosette are usually very small or absent. Leaves may be chlorotic, although no distinct pattern is consistently seen.

Manganese (Mn)

Manganese deficiency causes chlorosis of leaf tissue, beginning at the leaf margin and progressing inward (Plate 244). Tissue adjacent to leaf veins remains green, producing a pattern similar to that caused by Fe shortages. However, symptoms of Mn deficiency are usually seen first on older leaves (terminal leaves are first affected by Fe deficiency). A limited amount of chlorosis may not adversely affect the tree, but severe cases can completely stop growth (18).

Copper (Cu)

Symptoms of Cu deficiency have been described on apple (4,34) and pear (7,34). Necrotic lesions developing on the terminal leaves of rapidly growing shoots are the first indication of a Cu shortage. Leaves eventually abscise, and the terminal portion of the shoot may die. Branching occurs the following season from below the point of death, often resulting in witches'-broom growth. Trees appear "brushy" and low in vigor (Plate 245).

Boron (B)

Boron deficiencies have been reported on apple and pear in various regions of the world. In addition, B toxicity is a concern in some arid regions where levels in well water are excessive.

Vegetative symptoms of B deficiency include dieback of shoot tips. This most often occurs in late summer, but more severely deficient trees may exhibit symptoms earlier. The youngest leaves first turn slightly chlorotic, and margins may burn off. Branches may proliferate just behind the affected area, causing a witches'-broom growth that is subject to winter injury. Blossom blast is a specific disorder of pear caused by an early-season shortage of B (1) (Plate 246). Blossoms on affected trees begin to open normally but soon wither and die. Leaves on the same trees are usually not affected. The condition

develops sporadically and usually affects only a small percentage of trees in a given orchard. Boron applications may also improve fruit set on apple (13) and pear (3).

Fruit from B-deficient trees are often small, misshapen, and cracked (Plate 247). Several corking disorders of fruit (drought spot, internal cork, corky core, York spot, bitter pit) can be partly or completely eliminated by B applications (19). Lesions developing early in the season usually distort fruit shape severely, whereas those developing later may not change fruit shape. Boron deficiency may also induce symptoms similar to the "measles" disorder of bark caused by excessive accumulations of Mn (Plate 248).

Toxicity Symptoms

Relatively few nutrient toxicities are common in apple and pear orchards. Most toxicities result from excessive nutrient applications.

Nitrogen

Although N fertilizers rarely cause direct toxicity to apple or pear trees, excessive rates of N may have adverse effects on fruit quality and tree hardiness. Trees receiving too much N produce large, dark green leaves that remain on the tree late into the fall. Shoot growth continues into the autumn, and trees are more susceptible to winter injury. Excessive N may also delay fruit production on young trees and increase the susceptibility of apple (25) and pear (37) to fire blight.

Fruit on overfertilized trees are usually large, although they color poorly and mature later. Fruit flesh is less firm, and fruit deteriorate more rapidly after harvest or during storage. High N rates have reduced the soluble solids content and flavor of pears (30) and the color and firmness of apples (11). Some physiological disorders associated with low fruit Ca levels are aggravated by high N. They are bitter pit and cork spot in apples and corking and alfalfa greening in pear. Trees are typically more susceptible to fire blight infection.

Manganese

The "internal bark necrosis" or measles disorder of apple trees is a widespread problem caused by a toxic accumulation of Mn in the bark. Symptoms include blistering, cracking, and peeling of the bark on young twigs and branches (Plate 249). Similar symptoms may also be induced by B shortages. The first indication of the disorder is a slight elevation of the bark of twigs. Cambial tissue under affected bark is brown. Severely affected branches are low in vigor and often die back. This disorder is most common on acidic soils, and Red Delicious appears to be more susceptible to the problem than other cultivars. The problem is often encountered on old orchard sites where soil pH has been reduced by years of acidic N fertilizer applications. Young, replant trees show these symptoms. Leaf Mn levels greater than 300 ppm are often associated with this disorder.

Boron

Boron toxicity can be a problem if excessive amounts are applied or if orchards are irrigated with high-B water. Vegetative symptoms of excess B can include a dieback of shoot tips. A distinctive yellowing or necrosis may develop along the main veins of leaves. Affected leaves usually abscise. Apple trees receiving heavy applications of B produce fruit that mature early and break down in storage.

Arsenic (As)

Arsenic toxicity can occur in older fruit production regions where lead arsenate was used prior to the 1950s to control insect pests. Excessive As levels cause a reddening and scorching of leaf margins and reduced growth. Stone fruit are typically more sensitive to As toxicity than apples or pears.

Table 16.2. Critical concentrations of nutrient elements in leaves of apple and pear[a]

| Element | Concentration | | |
	Deficiency	Normal	Toxicity
Apple			
N, %	<1.7	2.0–2.5	...
P, %	<0.13	0.15–0.30	...
K, %	<1.0	1.2–1.9	...
Ca, %	<0.7	1.5–2.0	...
Mg, %	<0.2	0.3–0.7	...
S, %	...	0.19–0.27	...
B, ppm	<20	20–60	>140
Cl, ppm	...	NE[b]	...
Cu, ppm	...	NE	...
Fe, ppm	...	40–250	...
Mn, ppm	<25	25–150	>300
Mo, ppm	0.05	0.1–0.2	...
Zn, ppm	<15	15–200	...
Pear			
N, %	<1.8	1.8–2.6	...
P, %	<0.11	0.12–0.25	...
K, %	<0.7	1.0–2.0	...
Ca, %	<0.7	1.0–3.7	...
Mg, %	<0.25	0.25–0.90	...
S, %	<0.01	0.1–0.3	...
B, ppm	<15	20–60	...
Cl, ppm	...	NE	...
Cu, ppm	<5	6–20	...
Fe, ppm	...	100–800	...
Mn, ppm	<14	20–170	...
Mo, ppm	...	NE	...
Zn, ppm	<16	20–60	...

[a] Data from Gagnard (21), Huguet (26), and Shear and Faust (33).
[b] Not established.

Critical and Sufficient Tissue Nutrient Levels

Nutrient analyses of tissues have become an important diagnostic and management tool in apple and pear production. Leaf samples are collected during the middle of the growing season for nutrient analysis. Leaves should be selected from the midsection of current-season shoots. Although deficient, sufficient, and excessive levels in leaves may vary slightly by cultivar or production region, generally accepted levels for apples and pears were compiled from several sources (21,26,33) and summarized in Table 16.2. Because the concentrations of most nutrients change as leaves age, samples collected early or late in the season cannot be interpreted closely on the basis of values in Table 16.2.

Deficient and sufficient levels of most elements appear to be the same for all apple and pear varieties, with the exception of leaf N concentrations. The fruit of some apple varieties color poorly if excessive N is applied, and this is a serious problem if fruit are to be marketed fresh. Sensitive varieties, such as MacIntosh and Golden Delicious, may perform best when slightly low leaf N levels are maintained. Higher N levels can be maintained in most other varieties, particularly for processing apples.

The nutrient content of apple and pear fruit may influence the occurrence of some physiological disorders and how well fruit maintain quality during extended storage periods. As a result, efforts have been made to predict fruit quality based on mineral analyses of fruit. Calcium levels in fruit flesh are most useful, although fruit P, K, and Mg levels are occasionally considered as well.

Management Considerations

Apples and pears are high-value, intensively managed crops relative to many other commodities. Since even marginal nutrient deficiencies or toxicities can significantly reduce yield, fruit quality, and crop value, the nutritional status of apple and pear trees should be monitored closely.

Symptoms of nutrient deficiencies usually indicate an acute shortage. Since yield, fruit quality, or tree vigor are often adversely affected by the time symptoms become apparent, visual assessment of tree appearance is not a sufficient basis for fertilization decisions. Periodic soil testing provides critical information on pH changes and general nutrient supplies. However, soil nutrient levels often correlate poorly with tree nutrient status, so soil tests only estimate fertility requirements.

Nutrient deficiencies and excesses are most easily diagnosed in apples and pears by leaf analysis. Leaf analysis also provides a means of avoiding potential problems, since nutrients approaching deficient levels can be identified before yield or fruit quality are adversely affected.

Selected References

1. Batjer, L. P., and Rogers, B. L. 1953. "Blossom blast" of pears: An incipient boron deficiency. Proc. Am. Soc. Hortic. Sci. 62:119-122.

2. Batjer, L. P., Rogers, B. L., and Thompson, A. H. 1952. Fertilizer applications as related to nitrogen, phosphorus, potassium, calcium and magnesium nutrition by apple trees. Proc. Am. Soc. Hortic. Sci. 60:1-6.

3. Batjer, L. P., and Thompson, A. H. 1949. Effect of boric acid sprays applied during bloom upon the set of pear fruit. Proc. Am. Soc. Hortic. Sci. 53:141-142.

4. Benson, N. R. 1962. Some new minor element problems in Central Washington orchards. Wash. State Hortic. Assoc. pp. 19-22.

5. Benson, N. R., Woodbridge, C. G., and Bartram, R. D. 1970. Nutrient disorders in tree fruit. Wash. State Univ. Coop. Ext. Serv. Pac. Northwest Bull. 121.

6. Blake, M. A., Nightingale, G. T., and Davidson, O. W. 1937. Nutrition of apple trees. New Jersey Agric. Exp. Stn. Bull. 626.

7. Bould, C. D., Nicholas, J. D., Potter, J. M. S., Tolhurst, J. A. H., and Wallace, W. 1950. Zn and Cu deficiency of fruit trees. Bristol Univ. Ann. Rep. Agric. Hortic. Res. Stn. Long Ashton, pp. 45-49.

8. Bould, C., and Parfitt, R. I. 1973. Leaf analysis as a guide to the nutrition of fruit crops. X. Magnesium and phosphorus sand culture experiments with apple. J. Sci. Food Agric. 24:175-185.

9. Boynton, D., and Burrell, A. B. 1944. Potassium-induced magnesium deficiency in the McIntosh apple tree. Soil Sci. 58:441-454.

10. Boynton, D., Cain, J. C., and Van Geluwe, J. 1943. Incipient magnesium deficiency in some New York apple orchards. Proc. Am. Soc. Hortic. Sci. 42:95-100.

11. Boynton, D., and Oberly, G. H. 1966. Apple nutrition. Pages 1-50 in: Nutrition of Fruit Crops. N. F. Childers, ed. Horticultural Publications, Rutgers University, New Brunswick, NJ.

12. Boynton, D., and Oberly, G. H. 1966. Pear nutrition. Pages 489-503 in: Nutrition of Fruit Crops. N. F. Childers, ed. Horticultural Publications, Rutgers University, New Brunswick, NJ.

13. Bramlage, W. J., and Thompson, A. H. 1962. The effects of early season sprays of boron on fruit set, color, finish and storage life of apples. Proc. Am. Soc. Hortic. Sci. 80:64-72.

14. Bryant, L. R., and Gardner, R. 1943. Phosphorus deficiency in pears. Proc. Am. Soc. Hortic. Sci. 42:101-103.

15. Burrell, A. B., and Boynton, D. 1943. Response of apple trees to potash in the Champlain Valley III. Proc. Am. Soc. Hortic. Sci. 42:61-64.

16. Burrell, A. B., and Cain, J. C. 1941. A response of apple trees to potash in the Champlain Valley of New York. Proc. Am. Soc. Hortic. Sci. 38:1-7.

17. Childers, N. F. 1966. Nutrition of fruit crops. Horticultural Publications, Rutgers University, New Brunswick, NJ.

18. Dunne, T. C. 1946. "Withertip" of apple trees. J. West. Austr. Dep. Agric. 23:127-130.

19. Faust, M., and Shear, C. B. 1968. Corking disorders of apples: A physiological review. Bot. Rev. 34:441-469.

20. Fisher, E. G., and Kwong, S. S. 1961. The effect of potassium fertilization on quality of McIntosh apple. Proc. Am. Soc. Hortic. Sci. 78:16-23.

21. Gagnard, J. 1984. Apples. Pages 207-209 in: Plant Analysis as a Guide to the Nutrient Requirements of Temperate and Tropical Crops. P. Martin-Prevel, J. Gagnard, and P. Gautier, eds. Lavoisier Publishing, New York.

22. Greenham, D. W. P. 1976. The fertilizer requirements of fruit trees. Proc. Fertil. Soc. Lond. 157:32.

23. Harley, C. P. 1947. Mg deficiency in Kieffer pear trees. Proc. Am. Soc. Hortic. Sci. 50:21-22.

24. Haynes, R. J., and Goh, K. M. 1980. Distribution and budget of nutrients in a commercial apple orchard. Plant and Soil 56:445-457.

25. Hildebrand, E. M., and Heinicke, A. J. 1937. Incidence of fire blight in young apple trees in relation to orchard practices. Cornell Agric. Exp. Stn. Mem. 203.

26. Huguet, C. 1984. Apples. Pages 230-248 in: Plant Analysis as a Guide to the Nutrient Requirements of Temperate and Tropical Crops. P. Martin-Prevel, J. Gagnard, and P. Gautier, eds. Lavoisier Publishing, New York.

27. Moon, H. H., Harley, C. P., and Regeimbal, L. O. 1952. Early-season symptoms of magnesium deficiency in apple. Proc. Am. Soc. Hortic. Sci. 61:61-65.

28. Mulder, D. 1950. Mg deficiency in fruit trees on sandy soils and clay soils in Holland. Plant Soil 2:145-157.

29. Raese, J. T. 1977. Response of young 'd'Anjou' pear trees to triazine and triazole herbicides and nitrogen. J. Am. Soc. Hortic. Sci. 102:215-218.

30. Raese, J. T. 1982. Response of Anjou pear trees to nitrogen fertilizer and other elements in the Pacific Northwest. In: The Pear, Cultivars to Marketing. T. van der Zwet and N. F. Childers, eds. Horticultural Publications, Gainesville, FL.

31. Raese, T. 1987. Delicious apple trees respond to phosphate fertilization. Goodfruit Grower, April, pp. 34-39.

32. Shear, C. B. 1975. Calcium-related disorders of fruits and vegetables. HortScience 10:361-365.

33. Shear, C. B., and Faust, M. 1980. Nutritional ranges in deciduous tree fruits and nuts. Hortic. Rev. 2:142-164.

34. Smith, R. E., and Thomas, H. E. 1928. Copper sulfate as a remedy for exanthema in prunes, apples, pear and olives. Phytopathology 18:449-454.

35. Southwick, L. 1943. Magnesium deficiency in Massachusetts apple orchards. Proc. Am. Soc. Hortic. Sci. 42:85-94.

36. Toenjes, W. 1947. Occurrence of fireblight in Bartlett pears as influenced by certain cultural practices. Mich. Agric. Exp. Stn. Quart. Bull. 32:143-148.

37. Toenjes, W. 1949. Occurrence of fireblight in Bartlett pears as influenced by certain cultural practices. Mich. Agric. Exp. Stn. Quart. Bull. 32:143-148.

38. Van Slyke, L. L., Taylor, O. M., and Andrews, W. H. 1905. Plant food constituents used by bearing fruit trees. New York Agric. Exp. Stn. Bull. 265:205-223.

39. Wallace, T. 1928. Leaf scorch in fruit trees. J. Pomol. Hortic. Sci. 7:1-31.

40. Wallace, T. 1951. The Diagnosis of Mineral Deficiencies in Plants by Visual Symptoms. A Color Atlas and Guide. His Majesty's Stationary Office, London.

Chapter 17

Heinz K. Wutscher
USDA Horticulture Research Lab
Orlando, Florida

Paul F. Smith
USDA Horticulture Research Lab
Orlando, Florida

Citrus

Citrus is grown on a wide range of soils and under varied climates in the belt between 35° north and south latitudes. In the true tropics, citrus production is mostly on a small scale for local consumption; commercial production is concentrated in the subtropical areas. Citrus trees are usually budded onto rootstocks, which may have a profound influence on mineral uptake, drought resistance, disease susceptibility, productivity, and fruit quality. More than with other tree crops, rootstocks have been used in citrus to overcome salinity, high pH, drainage, and other soil-related problems. Some of the most successful rootstocks are from related genera or intergeneric hybrids.

Nutrient balance is an important factor in the appearance of deficiency symptoms because deficiencies may be caused by an excess of one element rather than a shortage of another. Deficiencies of some nutrients cause stress symptoms in young leaves. This is true for boron, calcium, copper, zinc, manganese, and iron. Other elements induce symptoms in leaves of any age. Such is the case with nitrogen and sulfur. Magnesium, potassium, and molybdenum show visible symptoms mostly in mature leaves. In the more developed areas of the world, citrus fertilization technology has reached a high degree of sophistication, mostly on the basis of leaf analysis. This permits remedial action before visible stress symptoms develop. Still, visual symptoms of nutritional disorders of citrus remain an important warning signal of problems, and their recognition remains an important facet of citrus management. Deficiency symptoms frequently occur first on only a few trees or even only on a few shoots in an entire orchard, thus giving a warning of potentially serious deficiency in the future.

Numerous varieties of citrus are cultivated either for fruit production or for ornamental purposes. They range from small shrubs to large trees, from evergreen to deciduous, and from simple to compound leaves. Stress symptoms are sufficiently similar across this wide range of citrus types and under different climatic conditions that a skilled observer can reach a correct diagnosis most of the time.

There is no exact relationship between the concentration of each essential element and plant function. For most elements, a rather wide range in concentration is compatible with good growth and high productivity. Table 17.1 illustrates the ranges found for oranges grown in both subtropical and desert climates.

Deficiency Symptoms

Nitrogen (N)

Nitrogen has greater influence on growth and yield than any other nutrient element, which is indicated by the narrow spread in tolerance of satisfactory leaf concentration listed in Table 17.1. It is a constituent of amino acids, amides, enzymes, alkaloids, chlorophyll, and other compounds important in plant metabolism, especially the various proteins. It is absorbed in either the nitrate or ammonium form. Most of the ammonium is incorporated into organic compounds in the roots, whereas nitrate can be translocated and then reduced in other parts of the plant. Nitrogen is highly mobile, but most of the N used in growth flushes is translocated from older tissues rather than derived from current absorption. Increases in leaf N can be detected in as little as 5–7 days after soil application.

Nitrogen deficiency has many manifestations. Under continuous stress for N, the trees maintain substantial growth by recycling N from older to newer leaves. Thus, the life of a leaf may be shortened from 1–3 years to 6 months or so. As the mature leaf loses N to younger leaves, it turns yellow and abscises, leaving only young, pale green, terminal leaves on the tree (Plate 250). Thinness of foliage is very striking as the inside leaves are lost prematurely. Such trees suffer from exposure and are prone to sunscald and freeze injury. Twigs may die back and fruit production be limited to small crops of inferior quality.

Under only mild stress for N, the premature shedding of old leaves is reduced, and leaf duration on the tree may extend to 12–18 months before abscission. Recycling of N is still evident, as the old leaves yellow just before abscission (Plate 251). This level of N is usually indicated by a 2.0–2.3% content in 5- to 7-month-old leaves. Fruit production and quality may be satisfactory under some conditions, but maximum fruit set and yield require a higher N level. A tree that is adequately supplied with N (2.5–3.0% in leaves) will sustain several successive cycles of leaf growth, often up to 2 or 3 years each, before abscission takes place. The old leaves do not turn yellow before dropping. Such trees have a dense canopy that protects both tree and fruit from weather damage.

Toxicity or ill effects of excess N on fruit production are virtually unknown short of actual salt burn. Nitrogen rates of 100–300 lb/acre (110–335 kg/ha) per year are commonly used. The quantity depends on tree age, production potential, density of planting, etc. The higher rates are beneficial for fruit grown for processing, because they induce higher yields of juice and soluble solids along with more intense juice color. This is accompanied by smaller fruit with thinner rinds. For fresh fruit, somewhat lower rates are generally favored, as this tends to induce larger fruit and retard the "regreening" of the peel during spring and summer. Regreening is a problem in late-maturing varieties where the fruit may remain on the tree for 12–18 months.

Citrus will utilize N from many sources; the most common are ammonium nitrate, urea, ammonium sulfate, and calcium nitrate. Sodium nitrate and manures are also used to some extent.

Table 17.1. Critical concentrations[a] of nutrients in citrus leaves

Element	Deficient[b]	Satisfactory	Excess[c]
N, %	<2.5	2.5–2.8	>2.8
P, %	<0.10	0.1–0.17	>0.17
K, %	<0.8	0.8–1.7	>1.7
Ca, %	<2.6	2.6–5.0	>5.0
Mg, %	<0.19	0.19–0.5	>0.5
S, %	<0.2	0.2–0.5	>0.5
Cl, %	. . .	0.2–0.4	>0.4
Na, %	. . .	0.01–0.15	>0.25
Fe, ppm	<35	35–130	>130
Zn, ppm	<19	19–50	>50
Mn, ppm	<19	19–100	>100
Cu, ppm	<5	5–15	>15
B, ppm	<25	25–200	>200
Mo, ppm	<0.07	0.07–0.25	>0.3

[a] Concentrations are for 4- to 6-month-old leaves on nonfruiting twigs on a dry weight basis.
[b] For deficiency, visible symptoms usually require even lower levels than those given.
[c] For excess, it is unlikely that greater amounts will be beneficial but toxicity symptoms occur only at higher values.

Sodium nitrate can induce sodium toxicity at high rates, and its extended use on clay soils may lead to water penetration problems. Contamination of urea with biuret and potassium nitrate with perchlorate may induce chlorosis.

Phosphorus (P)

Phosphorus is not reduced in plants but remains in its highest oxidized form. Its most important functions are as a constituent of nucleic acids in DNA and RNA molecules and of phospholipids in membranes and also as a participant in energy transfer and in the regulation of carbohydrate metabolism. Phosphorus compounds are readily translocated and concentrate in metabolically active tissues.

Phosphorus deficiency is rarely found in citrus anywhere in the world. Even soils with low total P content generally provide adequate amounts for intensively grown citrus. In Florida, isolated spots of white sand and others of pure muck have induced deficiency symptoms. In California, a few cases have been found on heavy clays. Visible leaf symptoms occur only at leaf levels below 0.07%, and a percentage this low is rarely encountered.

Deficiency symptoms include reduced growth and yields, premature defoliation, and reduced size of new leaves that develop a bronze cast. Fruit quality is markedly affected by low P levels. Coarse, thick peels, pale color, spongy fruit, and excessive acidity of juice may occur (Plate 252).

Toxicity of any form of P is unknown in citrus, although excessive P is known to induce disturbances to the nutrition of several micronutrients such as Zn and Fe.

Phosphorus deficiency on light sandy or muck soils can be readily corrected in a few months by soil application. On heavy soils, surface applications may not be effective, and soil injection may be required to get P into the root zone. Superphosphate, treble superphosphate, or ammonium phosphate are the main sources of P.

Potassium (K)

Potassium is highly mobile both within the cell and through the xylem and phloem. It tends to concentrate in the cytoplasm and vacuoles. It functions in buffering anions and stabilization of internal pH, enzyme activation, protein synthesis, photosynthesis, in turgor-related processes such as cell extension, stomatal functions, nyctinastic and seismonastic movements, and in transport of metabolites.

Citrus trees perform normally over a wide range of leaf K levels, but visible deficiency symptoms tend to appear when the leaf level drops below 0.3–0.4%. Weak new growth and twig dieback as well as small crinkled leaves that may develop yellow-to-bronze chlorosis in the outer half of the leaf may result from prolonged

K stress (Plate 253). In contrast to N and Mg deficiencies, which induce premature shedding, the leaves tend to be persistent. After extended periods of time, the chlorotic leaf tips turn brown, giving an appearance of leaf scorch or "tipburn."

Potassium has more effect on fruit quality than any of the other elements. High levels induce large, coarse fruit with thick peels, greenish color, and juice with high acidity. As the K level is reduced, fruit size lessens (Plate 254), peels thin, fruit color improves, juice acidity decreases, and juice percentage increases. Commercially grown fruit is usually from trees maintained with leaf K between 1.0 and 2.0%, but leaf levels up to 4 or 5% are not detrimental to the tree. Potassium is antagonistic to the absorption of Mg, Ca, and ammoniacal N. While no toxic symptoms occur with high rates of K, Mg deficiency is induced, and leaf levels of Ca and N are lowered.

Potassium chloride and sulfate are the most commonly used sources of K. For soil applications, use rates similar to those for N (100–250 lb/acre [110–300 kg/ha] of K_2O per year). Foliar sprays of potassium nitrate are sometimes effective where heavy soils limit the penetration of K to the root zone; however, repeated sprays are required.

Calcium (Ca)

Calcium is a relatively large divalent cation that readily enters the apoplast and is bound in exchangeable form to cell walls and the exterior surface of the plasma membrane. It occurs only in low levels in the cytoplasm and is apparently not very active in metabolic processes. Its mobility from cell to cell and in the phloem is very low. Nontoxic even at high concentrations, it serves as a detoxifying agent, tying up toxic compounds and maintaining the cation-anion balance in the vacuole. In the cell walls, Ca serves as a binding agent in the form of calcium pectates.

Calcium is generally the dominant nutrient in citrus leaves. Normal growth proceeds over a wide range of leaf Ca. Concentrations ranging from 1.5 to 7.0% have been found in seemingly healthy trees. Calcium deficiency symptoms are extremely rare. Under Ca stress, strong substitution of other base elements (Na, K, Mg) occurs, and excess H^+ abounds. Leaf levels of less than 1% Ca are associated with stunted growth, poor root development, dieback of twigs, and yellowing of leaf margins and interveinal areas (Plate 255). Calcium concentration in the leaves increases with age, and Ca is precipitated mainly as oxalate. Iron deficiency ("lime-induced chlorosis") on high-Ca soils is a commonly observed effect. In some cases, the anion associated with Ca (Cl^-, SO_4^{2-}) becomes toxic, rather than Ca itself.

Soils vary widely in their Ca content. Those low in Ca usually are also low in pH, and when lime (calcium carbonate) is applied to raise the pH, an ample supply of Ca is provided. Other sources of Ca are the irrigation water, especially from deep wells, fertilizers such as calcium nitrate and superphosphate, and fillers used in other fertilizers.

Magnesium (Mg)

The best known function of Mg is its role as the central atom of the chlorophyll molecule. Only a small part of the Mg in the plant is bound to chlorophyll, however. It performs a key role in the regulation of cellular pH and the cation-anion balance. Magnesium has an essential function as a bridging element for the aggregation of ribosome subunits in protein synthesis. Many enzyme reactions (e.g., phosphatases, ATPases, carboxylases) require Mg. Other cations, such as K, NH_4, Ca, and Mn as well as H, can depress Mg uptake.

Magnesium is readily translocated from older to younger tissues and fruit; therefore, Mg deficiency symptoms appear primarily on mature leaves containing less than 0.2% Mg. The leaves become "bronzed," and yellowish blotches appear between the main veins and along the midrib (Plate 256). The margins of the leaves fade inward, leaving an inverted, green V at the leaf base.

Magnesium deficiency is corrected with soil application of magnesium sulfate or dolomite. On acid soils, pH maintenance with dolomitic lime generally provides adequate Mg for the tree and eliminates the need for soluble Mg salts. Repeated magnesium nitrate sprays may prevent symptom development.

Sulfur (S)

Sulfur is a constituent of the amino acids cystine, cysteine, and methionine and therefore of proteins. Sulfur deficiency inhibits the synthesis of ferredoxin, biotin, and thiamine pyrophosphate. Sulfur can be taken up as SO_2 by the aerial parts of plants, but it is mostly absorbed as the divalent anion SO_4^{2-} by the roots. Sulfur is highly mobile in the plant, and the S content of citrus leaves changes little with age.

Sulfur deficiency symptoms in citrus are similar to those of N deficiency and occur only when the S level in the leaves falls below 0.19%. These symptoms have only been observed on plants in solution culture.

Sulfur exists in the soil as free S and sulfate and in diverse organic and inorganic compounds. Industrial air pollution is a major source of S; others are sulfate-containing compounds such as magnesium sulfate, ammonium sulfate, and gypsum. Elemental S is often used as a pesticide. Sulfur is also applied in most irrigation water.

Sodium (Na)

Sodium has been shown to be an essential minor element for some plants, but no benefit of Na has ever been shown for citrus.

Iron (Fe)

Iron is taken up mostly in the Fe^{2+} form. Transport is through the xylem in the form of complexes with citrate and peptide carbohydrate compounds. Two groups of Fe-containing proteins, the hemoproteins and the iron-sulfur proteins, are important in plant metabolism.

Iron deficiency, like Zn deficiency, occurs everywhere citrus is grown. The usual symptoms are very distinctive and consist of yellowed leaves with green midrib and lateral veins (Plate 257). This occurs only on young leaves but may persist into maturity without shortening the life of the leaf. In very mild cases, the symptoms may disappear gradually, or they may remain throughout the life of the leaf. In moderate to strong cases, the leaves may be solid yellow but persistent. In chronic cases, growth is restricted and dieback may occur. The symptoms may be atypical if commingled with Zn or Mn deficiencies.

Iron deficiency is of rare occurrence on neutral or acidic soils. White sands and some mucks may be deficient in Fe; they are readily corrected with occasional applications of a chelated source (FeEDTA or its hydroxyethyl form) at the rate of about 0.7 oz (20 g) of Fe per tree. Most Fe deficiency occurs on calcareous soils and is termed lime-induced chlorosis. This condition presents a very challenging problem. Foliar sprays of Fe are only partially effective at best, and massive chelate applications are required for soil treatment, such as NaFeEDDHA (ethylenediame-di[o-hydroxyphenyl-acetic acid]) at the rate of about 1.4 oz (40 g) of Fe per tree. Even then soil application is not always completely satisfactory. Iron toxicity is not known to be a problem in citrus production.

Zinc (Zn)

Zinc is taken up mostly as the divalent cation Zn^{2+}. Zinc acts either as a metal component of enzymes or as a functional, structural, or regulatory cofactor of many enzymes. It is transported mostly in the xylem as the free cation or bound to organic acids.

Zinc deficiency occurs everywhere citrus is grown. It produces a distinctive yellow mottle on leaves that are reduced in size (Plate 258). In mild form, only a few twigs may have symptoms, and tree function is not seriously affected. When many shoots are affected, the leaves become very narrow and short and almost completely yellowed. Continued stress leads to defoliation and twig dieback, and the fruit are small, misshapen, and bleached to whitish color. A wide range of patterns may be found when severely patterned leaves obtain enough Zn to make a partial recovery. Small green spots appear on yellowed leaves where particles of Zn from dust or spray residue land on the leaf. A gradual supply of translocated Zn may lead to green bands along the leaf veins.

Zinc deficiency often accompanies other metal deficiencies. Zinc and Mn or Zn and Fe often coexist on the same shoots, and the resulting chlorosis may be baffling (Plate 259). Leaf analysis may be necessary to complete the diagnosis, but leaves with typical patterns for each element usually can be found to confirm the dual deficiencies. Trees suffering from pathogenic diseases, e.g., citrus blight and tristeza, may exhibit leaf symptoms for Zn deficiency, probably because of interference in translocation of Zn within the tree. Zn toxicity is not recognized. Citrus leaves can tolerate a 20-fold increase in concentration above the incipient deficient level of about 20 ppm.

Zinc is almost always applied in sprays as ZnO or $ZnSO_4$ using 5 lb of Zn/500 gal of water (1.2 kg/1,000 L). It is, however, readily absorbed from acid soils if it is incorporated to depths where the roots can contact it.

Manganese (Mn)

Manganese is absorbed mainly as Mn^{2+} and translocated as the free divalent cation in the xylem. Its main function appears to be as a structural constituent of metalloproteins, where it acts as a binding site or as a redox system (Mn^{2+}/Mn^{3+}). It activates a number of enzymes, especially decarboxylases and dehydrogenases.

Manganese deficiency is widespread, but often is only of transient occurrence. Like the other heavy metals, Mn deficiency only occurs in young, enlarging leaves. In spring, when the soil is still cool, and in summer, when leaf expansion is very rapid, temporary Mn deficiency patterns are frequently found. Mostly they go away without any attention. Only persistent and severe pattern development of most of the foliage requires correction.

Manganese deficiency shows a light green mottle on a dark green background (Plate 260). Often the pattern appears like a series of inverted "horseshoes" around the periphery of the leaf. At times it may be a complete marbling of light green on dark green. Green-vein effects are either lacking or appear fuzzy in contrast to the sharp delineation of Fe deficiency. Leaf size is not reduced. If the stress of deficiency continues, the pattern persists into maturity and old age of the leaf until every leaf on the tree is patterned. Yellowish, bronzish colored patterns may indicate the imposition of another factor such as Zn or Mg deficiency or anion toxicity.

Manganese toxicity is unknown in field-grown citrus. On the other hand, it is difficult to induce Mn absorption from most soils. Unneutralized $MnSO_4$ at a rate of 7–12 lb of Mn/acre (8–14 kg/ha) per year is generally included in a postbloom pesticidal spray in Florida. In humid climates, finely ground MnO is very effective in sprays.

Copper (Cu)

Copper is present in three types of proteins: 1) blue proteins, which function in one-electron transfer such as in plastocyanin; 2) nonblue proteins, which produce peroxidases and oxidize monophenols; and 3) multi-copper proteins containing at least four Cu atoms per molecule, which act as oxidases, e.g., ascorbate oxidase and laccase.

Copper deficiency, unlike that of other heavy metals, is not associated with a leaf chlorosis. Abnormally large, dark green leaves give early indications of deficiency (Plate 261). Twigs die back, and multiple buds form near the juncture with live wood. The most reliable symptoms for identification of copper deficiency are gum pockets under the green bark of young wood (Plate 262) and brownish excrescences on fruit, twigs, and leaves. New shoots may produce narrow, elongated leaves similar to peach leaves. If the condition is moderate, fruit development is normal; with more severe deficiency, fruit develops brown lesions, the rigid peel may split, and the fruit drops.

Copper deficiency was first cured accidentally in Florida 75 years ago by throwing an embalming fluid (copper sulfate) around trees with severe "dieback." The spectacular response led to the inclusion of "bluestone" in nearly all citrus fertilizers for several decades until severe Cu toxicity developed in many Florida groves. Soil applications are no longer used even on new land, because fungicidal sprays more than adequately supply the nutritional Cu requirements (3.75 lb of Cu/500 gal of water; 0.9 kg/1,000 L). Either copper oxide or tribasic copper sulfate are extensively used in sprays. Copper toxicity damages the root system and leads to droughty trees with sparse foliage, very small leaves, and stubby feeder roots. Iron deficiency frequently occurs. Raising the soil pH to above 7 with high rates of lime is effective in reducing Cu availability and allowing tree recovery. If liming is neglected and the soil again becomes acidic, the trees will decline again.

Boron (B)

Boron occurs as the very weak acid H_3BO_3 in the soil solution and is taken up in this form by the roots. Transport from the roots to the shoots is confined to the xylem. Boron's role in plant nutrition is still only partially understood. Its functions appear to be mostly extracellular and related to lignification and xylem differentiation. Boron deficiency apparently leads to a build up of indoleacetic acid (IAA) by blocking IAA oxidase in the roots, which inhibits elongation growth. RNA and protein synthesis are inhibited. Membrane stability, pollen germination, and pollen tube growth are decreased. Both B deficiency and toxicity are widespread problems in citrus.

Boron deficiency causes only vague symptoms in the foliage. Short internodes, drab color, yellow veins similar to girdling, thickened leaves, corky venation, and buckling of the leaf blade along the midrib have all been associated with B deficiency but are not always easy to recognize in mild stages (Plate 263). Continued stress for B leads to a collapse of the phloem, which in effect is an internal girdling that produces symptoms similar to mechanical girdling or girdling produced by diseases (such as foot rot from *Phytophthora* fungus or phloem disruption by tristeza virus). The most definite symptom of deficiency is in the fruit. Gum deposits in the peel are rather specific (Plate 264). Bloom may be excessive, but unfruitfulness results from excessive shedding of young fruit.

Most soils or irrigation water in arid regions contain sufficient or even excessive B. In tropical and subtropical areas, boron leaches readily with high rainfall, and fertilizer or spray application are generally beneficial. In Florida, virtually all citrus receives one application each year in very small amounts. Rates of 1–3 lb/acre (1–3 kg/ha) of B may be applied either in the fertilizer or in a spray from a soluble borate salt.

Chlorine (Cl)

Chlorine is taken up as the chloride (Cl^-) ion, and it is highly mobile within the plant. Chlorine functions mainly in processes related to charge compensation and osmoregulation. The Hill reaction in photosynthesis requires Cl^-. Chloride, together with K^+, regulates stomatal opening. Chloride deficiency has never been demonstrated in citrus.

Molybdenum (Mo)

Molybdenum is an unusual nutrient element in several ways. It is required in smaller amounts than other essential elements, and its availability in the soil increases with pH. Although Mo is a metal, it occurs in aqueous solution mainly as molybdate oxyanion, MoO_4^{2-}, in its highest oxidized form. Its properties resemble those of nonmetals and other divalent anions.

Molybdenum deficiency causes a "yellow spot" to develop on fully matured leaves. Leaves produced in the spring develop one to several large yellow-orange spots in late summer (Plate 265). The spots occur at random on the leaf and gradually become gum-impregnated, turn reddish brown, and bulge on the underside of the leaf. During the fall, the leaves drop prematurely, leaving a thinly foliated tree.

Molybdenum deficiency appears to be a Florida phenomenon. For over 60 years, the symptoms were fairly widespread, but the cause was unknown until 1951. Sprays with sodium molybdate (0.8 oz/100 gal of water; 6 g/100 L) were found to prevent the symptoms. However, at about the same time a liming program was instituted to combat Cu toxicity. As a rule, the soil pH was raised from about 5.0 to 6.5, and in high-Cu groves to above 7. Molybdenum deficiency vanished as pH increased and is now a forgotten event. Molybdenum toxicity is not a problem in citrus.

Toxicity Symptoms

Toxicities caused by excesses of most essential elements are not too common in citrus. Those nutrients associated with saline and sodic problems—Na, Cl, and S—may often cause toxicity, but it results from the saline or sodic conditions in the soil (Plate 266). Boron can also cause toxicity. It occurs mainly in areas with high B levels in irrigation water.

Sodium

Sodium toxicity is of major importance in arid and semiarid areas or where Na-containing fertilizers or ground water are used. Symptoms of Na toxicity are irregular yellow mottling and tip and marginal leaf necrosis (Plate 267). Heavy leaf drop is often the first sign of Na toxicity.

Chlorine

Chloride toxicity is a major production problem of citrus, especially in dry or coastal areas. Chlorides accumulate in the leaves, and saline water sprayed on the leaves increases the chances of Cl^- buildup. The toxicity symptoms are a bronzing of the leaves, followed by tipburn, marginal necrosis, and defoliation (Plate 268).

Chloride toxicity can be readily eliminated when nonsaline water is applied and the soil leached with excess water. Rootstocks have been widely used in citrus to avoid Cl toxicity. Trifoliate and trifoliate hybrid rootstocks are particularly chloride sensitive, but rootstocks of Cleopatra and other mandarins and sour orange are chloride tolerant.

Sulfur

Sulfur is often associated with saline soil conditions. When S concentration in the leaves exceeds 0.5%, a mottling of leaves occurs. Sulfur toxicity exists but is normally less serious than that of Cl or Na.

Boron

Boron toxicity symptoms appear in leaves with greater than about 250 ppm of B and in dry areas where the irrigation water contains more than 0.5 ppm of B. There is a yellowing of the leaf tips and mottling, followed by the formation of brownish, resinous pustules on the undersides of leaves and premature leaf drop (Plate 269). Unless the toxicity is severe, there is no adverse effect on the fruit.

Other Toxicities

The effects of Cu toxicity were discussed under Cu deficiency symptoms. Excesses of certain other nutrients can cause imbalances. Excess soil P often induces deficiencies of Zn and Fe, and excess P in the plant can induce nutritional imbalances of Fe and Zn. High rates of K may induce a Mg deficiency, and leaf levels of Ca and N are lowered.

Selected References

1. Chapman, H. D. 1966. Diagnostic Criteria for Plants and Soils. Citrus Research Center, University of California, Division of Agricultural Sciences, Riverside.
2. Chapman, H. D. 1968. The mineral nutrition of citrus. Pages 127-289 in: The Citrus Industry. Vol. 2, Anatomy, Physiology, Genetics and Reproduction. W. Reuther, L. D. Batchelor, and H. J. Webber, eds. University of California, Division of Agricultural Sciences, Berkeley.
3. Embleton, T. W., Reitz, H. J., and Jones, W. W. 1973. Citrus fertilization. Pages 122-182 in: The Citrus Industry. Vol. 3, Production Technology. W. Reuther, ed. University of California, Division of Agricultural Sciences, Berkeley.
4. Jones, W. W., and Smith, P. F. 1964. Nutrient deficiencies in citrus. Pages 359-414 in: Hunger Signs in Crops: A Symposium. 3rd ed. H. B. Sprague, ed. David McKay, New York.
5. Koo, R. C. J. 1984. Recommended fertilizers and nutritional sprays for citrus. Univ. Fla. Agric. Exp. Stn. Bull. 536D.
6. Marschner, H. 1986. Mineral nutrition of higher plants. Academic Press, New York.
7. Pratt, R. M. 1958. Florida Guide to Citrus Insects, Diseases and Nutritional Disorders in Color. University of Florida Agricultural Experiment Station, Gainesville.
8. Smith, P. F. 1966. Citrus nutrition. Pages 174-206 in: Temperate to Tropical Fruit Nutrition. N. F. Childers, ed. Horticultural Publications, Rutgers University, New Brunswick, NJ.

Chapter 18

R. Scott Johnson
Department of Pomology
University of California Kearney Agricultural Center, Parlier

Stone Fruit: Peaches and Nectarines

Peaches are grown in many areas throughout the United States. Nectarines have mainly been grown in California in the past but are gaining in popularity in other areas as new cultivars adapted to those areas are developed. Peaches and nectarines are very vigorous trees and respond well to nitrogen fertilization. On good soils, nitrogen is often the only nutrient that needs to be added on a regular basis. On poorer or less fertile soils, deficiencies of potassium, magnesium, manganese, iron, zinc, or boron may develop. Deficiencies of phosphorus, calcium, sulfur, and copper are rarely seen.

According to a nutrient removal study by Rogers and co-workers (6), N and K are the two elements permanently removed in greatest quantity from a peach orchard (Table 18.1). This study was conducted over 35 years ago under moderate fertilization rates. The value for N could be even higher in many modern-day orchards, where yields are large and high rates of N fertilizer are used. Calcium is also found in large quantities in peach trees, but mostly in leaves and shoots, so it is eventually recycled. Very little P is permanently removed, sub-

stantially less than the amounts reported for many field crops. This could be part of the reason why peach trees have seldom been reported to be deficient in Ca or P. The quantity of Mg removed is also quite small.

Deficiency Symptoms

Nitrogen (N)

Nitrogen is the one nutrient that must be supplied on a regular basis to peach and nectarine trees wherever they are grown in the United States. This nutrient is needed to produce consistently high yields of large fruit.

Leaf symptoms range from a pale green to yellow over the whole leaf surface. Symptoms occur quite uniformly throughout the tree and along a shoot. Red and brown spots often develop on leaves (Plate 270) leading to a "shothole" appearance or a red tinge especially on basal leaves. Leaf size is reduced.

Shoot growth is markedly restricted. Shoots become spindly and have a very characteristic red coloration

Table 18.1. Amounts of nutrients removed from soil by 14-year-old peaches[a,b]

Plant part	Nitrogen		Phosphorus		Potassium		Calcium		Magnesium	
	lb/acre	kg/ha	lb/acre	kg/ha	lb/acre	kg/ha	lb/acre	kg/ha	lb/acre	kg/ha
Nutrients removed permanently										
Fruit	50.9	57.0	7.8	8.7	78.3	87.7	2.4	2.7	2.2	2.5
Top growth	7.0	7.8	0.8	0.9	2.2	2.5	7.0	7.8	0.8	0.9
Root growth	10.4	11.6	2.4	2.7	5.6	6.3	5.1	5.7	0.9	1.0
Subtotal	68.3	76.5	11.0	12.3	86.1	96.4	14.5	16.2	3.9	4.4
Removed from soil but ultimately returned										
Fruit drop and fruits thinned	19.2	21.5	2.4	2.7	21.6	24.2	1.9	2.1	1.1	1.2
Leaves	25.8	28.9	2.4	2.7	59.1	66.2	74.6	83.6	15.9	17.8
Prunings	16.2	18.1	1.9	2.1	8.6	9.6	21.6	24.2	1.9	2.1
Total	129.5	145.0	17.7	19.8	175.4	196.5	112.6	126.1	22.8	25.5

[a]Source: Rogers et al (6).
[b]On the basis of a yield of 15 tons/acre (33.6 t/ha).

(Plate 271). Leaf senescence occurs earlier in the fall and colors are brighter. Trees are more susceptible to winter cold injury, but heavily fertilized trees that are delayed from going into dormancy can also be more susceptible than more optimally fertilized trees.

The reduced shoot growth and decreased flowering tend to lower production. Fruit are smaller, even when trees are thinned to the same fruit load as a tree with better nutrition. However, fruit quality may be enhanced, showing a greater degree of redness. Fruit maturity is also advanced by N deficiency.

Phosphorus (P)

Phosphorus deficiency is rarely seen in commercial peach and nectarine orchards. It was first reported in the late 1940s in the Sandhills area of North Carolina but has not been reported in recent years. Part of the explanation for this may be that very little P is removed by the crop (about 5–10 lb/acre [6-11 kg/ha]), and senescing peach leaves are quite efficient at recycling P (about 70% recovery). Phosphorus deficiency is occasionally a problem in nursery stock where soil fumigation has killed the mycorrhizae that help in the uptake of P.

Leaf symptoms may start with a dark green color that eventually turns bronze or reddish purple. Leaves develop a very characteristic leathery texture. A red coloration appears on petioles and young shoots. Eventually leaf size is reduced, and premature defoliation may occur, beginning with basal leaves.

Yield and fruit size are reduced by P deficiency. The fruit is redder and maturity is advanced, but fruit quality is negatively affected. Fruit is disfigured by gumming, skins crack, and the flavor is very unpleasant.

Potassium (K)

Potassium is found in quite large quantities in both leaf and fruit tissues of peaches and nectarines. It appears to have an important role in fruit growth, since fruit size is markedly reduced by K deficiency. Deficiency is not uncommon and is encountered frequently on very sandy soils or in orchards where the topsoil has been scraped in a land-leveling operation. A heavy crop load can increase the likelihood of K deficiency.

Initial symptoms occur on midshoot leaves in early summer. Leaves are pale with a characteristic upward curling or rolling (Plate 272). Margins become chlorotic and then necrotic, leading to a marginal scorch. Affected leaves eventually develop cracks, tears, and necrotic spots.

Potassium-deficient trees have reduced flower bud formation and fruit size. Fruits are much less red or exhibit a dull, dirty looking, orange color and have poor storage quality.

Calcium (Ca)

Calcium deficiency in the field has only been briefly mentioned in one report from New Jersey. This is rather surprising considering the multitude of Ca-related disorders that have been reported for apples and other fruit trees. However, many of the problems occurring in apple fruits only show up after much longer periods of storage than those to which peaches and nectarines are generally subjected.

Where deficiency symptoms have been induced in sand culture, reduced shoot growth due to shortened internodes is observed. This is followed by defoliation and twig dieback. Leaf blades often roll inward and upward and may develop large chlorotic areas before they abscise.

Magnesium (Mg)

Magnesium deficiency has been reported from many places throughout the United States. It often occurs in sandy soils or those naturally low in Mg. It can also be induced by heavy applications of K-containing fertilizers, which replace the Mg on soil cation exchange sites.

Leaf symptoms start with a marginal interveinal chlorosis that may develop into necrosis. The center of the leaf stays green along the midrib creating an inverted "V" pattern (Plate 273). Basal leaves are most severely affected and generally abscise by late summer, leaving the bases of shoots bare. Until the deficiency becomes severe, shoot and fruit growth are not reduced, although early fruit drop may occur.

Sulfur (S)

Sulfur deficiency is extremely rare in peach-growing areas of the United States. This is partly because S is inadvertently applied in many different ways, such as in pesticides, fertilizers (ammonium sulfate), soil amendments (gypsum), manure, and other forms of organic matter. In industrial areas, atmospheric SO_2 levels can be sufficiently high to contribute a substantial amount of the S requirement. Sulfur is also efficiently recycled in senescing leaves. A few peach orchards in Washington State have reported S deficiency.

Sulfur deficiency in peach trees is characterized by uniformly yellow leaves at the shoot tip (Plate 274). Symptoms are similar to those of N deficiency except for location along the shoot; N-deficient plants are more uniformly affected along the shoot.

Iron (Fe)

Iron deficiency is quite common, especially in the western United States under high soil pH conditions. It often results from high levels of bicarbonate in the soil and is termed lime-induced chlorosis. Under these conditions, sufficient Fe is taken up, but the bicarbonate somehow inhibits its utilization by the tree.

The characteristic symptom of Fe deficiency in peach and nectarine is an interveinal chlorosis with the veins remaining distinctly green (Plate 275). As the severity increases, leaves can become almost white, eventually turning necrotic and abscising. Terminal leaves are affected before basal leaves, so some shoot dieback may occur.

Zinc (Zn)

Zinc deficiency is a fairly common nutrient disorder. It occurs throughout the United States but is especially a problem in California. It has often been called "little leaf," because rosettes of small pointed leaves form at shoot tips.

The first symptom of Zn deficiency is a chlorotic interveinal mottling that appears on older leaves. As the deficiency becomes more severe, internode and leaf growth at the shoot tip are greatly curtailed, leading to the little-leaf rosetting (Plate 276). Leaves often develop a very characteristic crinkled or wavy margin. Basal leaves eventually abscise, leaving tufts of small pointed leaves at the tip (Plate 277). Growth will generally be delayed the following spring, and extensive shoot dieback occurs. Flower bud formation is greatly reduced, and the few fruit that develop are small, misshapen, and of very poor quality.

Manganese (Mn)

Manganese deficiency is not common but has been reported in several different states. The characteristic symptom is an interveinal chlorosis extending from midrib to margin. Broad bands of green remain along the veins, giving a herringbone pattern (Plate 278). Symptoms tend to be general over the whole tree, although apical leaves often appear less affected. Severe deficiency causes some defoliation and shoot dieback. Moderate symptoms, however, don't seem to induce a reduction in shoot growth or yield. Temporary deficiency symptoms can occasionally be found in the spring under conditions of poor root growth induced by low soil temperatures.

Copper (Cu)

Copper deficiency is very rare in peach and nectarine orchards. Where it has been observed or induced in sand culture, symptoms include long narrow leaves with interveinal chlorosis (Plate 279). Early cessation of terminal growth and shoot dieback may occur. Often several buds start growing below the dead tip, leading to a "witches'-broom" effect.

Boron (B)

Boron deficiency is uncommon but has been reported in the northwestern United States, North Carolina, and British Columbia. Compared with other fruit trees, peaches are apparently less sensitive to B deficiency. The characteristic symptom of this disorder is a terminal dieback of twigs (Plate 280). New, weak growth is forced below these dead tips. Leaves that subsequently develop are small, thick, misshapen, and brittle. Flower buds

are particularly sensitive to B deficiency, so production is substantially reduced. Fruit develop dry, corky areas near the pit, and some cracking along the suture may occur.

Chlorine (Cl) and Molybdenum (Mo)

Both these elements are needed in only minute amounts in peach and nectarine trees. Deficiencies have never been reported in the field and to the knowledge of the author have never been induced in sand culture.

Toxicity Symptoms

Toxicity symptoms have been reported for the essential nutrients B and Cl. In addition, the nonessential elements sodium and arsenic can be toxic to peaches when supplied in large quantities.

Boron

Peach is one of the most sensitive fruit species to B toxicity. The disorder has been reported from several locations around the United States. Symptoms consist of shoot dieback at the tip with the bark becoming cracked and corky. Gum may exude from these cankers. Often several lateral shoots develop below the dieback. Leaf symptoms are generally confined to the midrib and main lateral veins. Flower bud formation and fruit set are reduced, and the resulting fruit are malformed, split, and have very poor flavor.

Chlorine and Sodium (Na)

Stone fruit in general are quite sensitive to an excess of Cl. This can be a particular problem in some irrigated orchards of the western United States where salts build up in the soil. Indications of moderate toxicity may be a reduction in vigor without any foliar symptoms. This may partly result from an osmotic effect in the soil. More severe toxicity results in a marginal leaf scorch (Plate 281) and eventual abscission of the leaf.

Leaf symptoms of Na toxicity are similar except the marginal burn is often striated. When Na is in excess in the soil, it tends to accumulate in the wood and roots rather than in the leaves. If the problem of excess salt is not corrected, tree death will result.

Arsenic (As)

Arsenic toxicity has been reported in peach trees planted on old apple orchard sites where arsenate sprays were used over many years. The first symptom of this disorder is a brown to red coloration along the margin of older leaves. Eventually spots of a similar color develop between the veins, turn necrotic, and drop out, giving a shothole appearance. Almost complete defoliation can occur, leaving fruit hanging on bare limbs. At maturity, the fruit is astringent to the taste.

Critical Tissue Concentration Levels

Diagnosing nutrient disorders can be greatly facilitated by using lab reports of leaf levels. The critical level for each element has been determined through research and experience (Table 18.2) and can help determine the severity or potential for a given deficiency or toxicity.

The concentration of most nutrients in peach leaves tends to level off in midsummer, so this time is used for sampling. Midshoot leaves are sampled from moderately vigorous shoots 8–20 in. (20–50 cm) in length located in well-exposed portions of the canopy. There can be substantial leaf-to-leaf variability in nutrient level; therefore, at least 80–100 leaves need to be sampled from one site. The leaves are then washed in a mild soap solution, rinsed, and oven-dried before being sent to the lab for chemical analysis.

The only exception to the time of sampling is for Fe. Often iron chlorosis does not result from insufficient Fe in the plant but from its inactivation by bicarbonate or some other chemical. Therefore, the leaf level may indicate a sufficient concentration. Taking leaf samples earlier in the season provides a better correlation with deficiency symptoms.

Management Considerations

A grower can have a significant effect on the nutritional status of his peach orchard through his management decisions. Certain deficiencies can be eliminated, nutrient uptake efficiencies can be improved, and environmental contamination can be reduced.

Balance of Nutrients

Applications of one fertilizer element to the soil can affect the uptake of other nutrients. This can occur through various ways including competition of cations for exchange sites and effects on nutrient solubility and on soil pH. Nutrient interactions have been demonstrated for many pairs of nutrients. From a management perspective, it is important to keep the various nutrients in balance. An excess of one element can often lead to deficiencies of other nutrients.

Nitrogen is one element that needs to be kept in balance. If trees become deficient in N, they lose vigor and productivity. However, excessive N can also lead to problems. The resulting shoot vigor leads to a dense canopy that delays fruit maturity, reduces the amount of red coloration on the fruit, shades out lower fruiting wood, and may lead to other fruit quality problems.

Soil and Orchard Floor Management

Many things can be done to the soil to either increase or reduce nutrient disorders. Land leveling where the topsoil is removed from one part of the field can lead to K deficiency. Potassium is generally concentrated in the top 6 in. (15 cm) of the soil and is very poorly supplied by the subsoil. Many management practices that lead to greater soil compaction or less soil aeration will cause a more rapid development of nutrient deficiency symptoms. Soil fumigation, which is often practiced before planting a new orchard, can have detrimental effects on the nutrition of the young tree. Fumigants that completely sterilize the soil will destroy the mycorrhizal association of fungi and roots that helps the tree take up such elements as P and Zn. Therefore, inoculation of mycorrhizal fungi or application of fertilizer may improve the nutritional status and early growth of the young tree.

Peach roots compete very poorly with weeds and grasses for water and nutrients. Therefore, young peach trees tend to have lower nutritional status, especially of N, and produce less growth when weeds are not controlled. This often happens even when extra N fertilizer is added. However, studies in the eastern United States have shown improved growth and nutritional status of young peach trees planted in killed sod compared with trees planted in bare soil. Cover crops in mature orchards can have the effect of either increasing or decreasing nutrient levels in leaves depending on how they are managed. In general, cover crops will out-compete the trees for nutrients. Therefore, higher fertilizer application rates will be needed to achieve the same nutrient levels in the tree. However, leguminous cover crops can supply a substantial amount of N to trees if they are incorporated at the correct time.

Irrigation Management

Both irrigation rate and application method can affect tree nutrition. Excess irrigation has been shown to lead to symptoms of Fe deficiency. This results from a lack of aeration in the soil that causes a buildup of carbon dioxide and bicarbonate, which leads to bicarbonate-induced Fe chlorosis. As long as the soil or irrigation water is not naturally high in bicarbonates, the problem is easily overcome with proper irrigation management.

Table 18.2. Critical levels of nutrients in peach and nectarine leaves taken in April or May

Nutrient	Deficient	Optimum range	Toxic
N, %	<2.3	2.6–3.0	...
P, %	...	0.1–0.3	...
K, %	<1.0	>1.2	...
Ca, %	...	>1.0	...
Mg, %	<0.25	>0.25	...
Cl, %	...	...	>0.3
Fe, ppm	<60	>60	...
Zn, ppm	<15	>20	...
Mn, ppm	<20	>20	...
Cu, ppm	...	>4	...
B, ppm	<18	20–80	>100
Mo, ppm	...	...	...

Overirrigation can also lead to reduced N uptake efficiency by leaching N below the root zone.

Nutrients applied through a low-volume drip or microsprinkler irrigation system have been shown to be taken up much more efficiently by fruit trees. For N, only about half as much fertilizer material is needed to achieve the same leaf levels as with broadcast fertilization and furrow irrigation. Iron chelates can also be applied at much lower rates through a low-volume irrigation system to overcome Fe deficiency. Potassium, which does not move readily through most soils, is more apt to do so when applied through a low-volume drip system.

Rootstock

Because nutrients are taken up by roots, it is natural to expect the rootstock to have an impact on tree nutrition. Some rootstocks are better than others at "foraging" for micronutrients. For example, Nemaguard rootstock, which is widely used in California, does not have a very fibrous root system and is rather poor at picking up micronutrients when the tree is young. Rootstocks vary considerably in their resistance to lime-induced chlorosis. Some of the newer peach-almond hybrids such as GF-677 are much more resistant to this disorder. Cultivars grown on their own roots without being grafted to a rootstock may have higher leaf levels of certain nutrients such as Ca.

Cultivar

Hundreds of peach and nectarine cultivars are grown commercially throughout the United States. Although they all have the same general requirements for nutrients, subtle differences exist in susceptibility to deficiencies, responses to fertilizers, and inherent vigor. In general, early maturing cultivars need less N than later cultivars. This is not only because fruit loads are lighter, and thus less N is removed in the crop, but also because vigorous shoot growth is more likely to occur once fruit is off the tree. Keeping the N level down helps prevent excessive shoot growth in early-season cultivars. Fruit color and other quality measures are much more adversely affected by overfertilization on some cultivars than others.

Other Cultural Practices

Several other cultural practices may influence tree nutritional status. During the autumn, many nutrients in leaves are remobilized and stored for growth the following spring in the branches and trunk and especially in the roots. Therefore, when potential growing points are removed by dormant pruning, the remaining growing points will have a proportionately greater supply of nutrients from the roots. This leads to higher leaf N levels in pruned than in unpruned trees.

Trunk or scaffold girdling is a cultural practice commonly used on early-season peach and nectarine cultivars. The practice increases fruit size and advances maturity but also causes chlorotic leaves that appear to be deficient in one or more nutrients. This condition, however, is rather temporary; once the girdling wound has healed, deficiency symptoms clear up, and the tree appears to be healthy.

Method of Fertilizer Application

To correct nutrient deficiencies, fertilizer materials can be applied to soil, foliage, or dormant wood, or they can be injected into the trunk depending on the nutrient involved. Nitrogen and K are typically needed in large enough quantities that they must be added to the soil. Potassium does not readily move through many soils and must, therefore, in many situations, be shanked to a depth of 6–8 in. (15–20 cm). A number of nutrient deficiencies including those of Mg, Mn, Zn, and B can be corrected with foliar applications. Zinc sprays can be sprayed on leaves during the season or on wood during dormancy. Correction of these deficiencies can also be achieved using soil-applied fertilizers, but rates are generally much higher.

Iron deficiency has been a particularly difficult disorder to correct. Foliar applications of Fe chelates can be effective, but correction is generally short lived. Iron chelates are also effective when applied to the soil but are generally needed at quite high rates. As mentioned previously, much lower rates can be used if fertilizer is applied to the soil through a low-volume irrigation system. Injecting various Fe-containing materials into the trunk has been demonstrated to work effectively; however, correction is still not permanent. To achieve permanent correction of the disorder, modification of the soil environment is required. Acidification of the whole soil profile using soil S is generally cost prohibitive, but some success has been achieved through acidifying a portion of the soil.

Selected References

1. Chapman, H. D. 1966. Diagnostic Criteria for Plants and Soils. Quality Printing, Abilene, TX.
2. Childers, N. F. 1966. Nutrition of Fruit Crops. Tropical, Sub-tropical, Temperate Tree and Small Fruits. Horticultural Publications, Rutgers-The State University, New Brunswick, NJ.
3. Childers, N. F., and Sherman, W. B., eds. 1965. The Peach. Horticultural Publications, Gainesville, FL.
4. Hambidge, G. 1941. Hunger Signs in Crops: A Symposium. Am. Soc. Agron. and Natl. Fert. Assoc., Madison, WI.
5. Johnson, R. S., and Uriu, K. 1989. Mineral Nutrition. Pages 68-81 in: Peaches, Plums, and Nectarines, Growing and Handling for Fresh Market. Publ. 331. University of California, Division of Agriculture and Natural Resources, Oakland.
6. Rogers, B. L., Batjer, L. P., and Billingsley, H. D. 1955. Fertilizer Applications as Related to Nitrogen, Phosphorus, Potassium, Calcium, and Magnesium Utilization by Peach Trees. Proc. Am. Soc. Hortic. Sci. 66:7-12.
7. Shear, C. B., and Faust, M. 1980. Nutritional ranges in deciduous tree fruits and nuts. Hortic. Rev. 2:142-163.

Chapter 19

Wilhelm Gärtel
Federal Biological Research Center
Institute for Plant Protection in Viticulture, Bernkastel-Kues, Germany

(Translated by H. O. Amberg, Clifton Springs, New York)

Grapes

Disorders in the nutrition of a grapevine manifest themselves by changes in the shape, color, chemical composition, performance, and attainable age of individual organs or the whole plant. The visible symptoms provide clues for their cause—the deficiency or excess of one or several nutrients. The appearance of the whole vine as well as of the vineyard may aid in the diagnosis. Soil analysis as well as that of certain parts of a plant, preferably leaves and petioles, can confirm nutrient imbalance (Table 19.1). These data, combined with a knowledge of the soil, the performance and sensitivity of the grape variety, and the cultural methods employed, complement the visual observation and increase the accuracy and significance of the diagnosis. Knowledge of the location of toxic emissions such as sulfur dioxide, nitrogen oxides, and fluoride (Plate 299) may also be helpful in areas where these situations occur.

From approximately 8,000 known *Vitis* cultivars, only a few are of economic importance in the world of viticulture. Grapevines adapt themselves to a wide range of soils and climates. They survive and sometimes produce profitable crops under conditions where other crops would fail. Intensive cultivation; an abundant, balanced supply of nutrients; and an integrated plant protection program may greatly increase production. The quantity of nutrients that grapevines take up during a vegetation period depends very much on soil, water supply, cultivar, rootstock, and yield. Tables 19.2 and 19.3 illustrate the average needs of essential elements and the annual nutrient removal of intensively cultivated grapevines.

Deficiency Symptoms

Nitrogen (N)

When N becomes limiting, the small, thin, stiff leaf blades turn pale green and then yellow. Young shoots, petioles, and cluster stems become pink or red, regardless of the cultivar (Plate 282). Under drought conditions, the leaf margins roll slightly upwards, wilt, and dry up. Sometimes light brown islands of dead tissue appear between the main veins of the basal leaves, and in extreme cases the pale leaf blades may wither and abscise. Nitrogen deficiency slows the growth of the shoots, and the berries start to mature early and do not reach their normal size. In commercial plantings, the first symptoms may appear after beginning of ripening, because N is

Table 19.1. Critical levels of nutrients in grapevine leaf blades[a,b]

Nutrient	Deficiency	Slight deficiency	Optimal level	Slight excess	Excess
N, %	<2.00	2.00–2.40	2.41–2.60	2.61–2.80	>2.80
P, %	<0.15	0.15–0.20	0.21–0.24	0.25–0.26	>0.26
K, %	<1.00	1.01–1.20	1.21–1.40	1.41–1.60	>1.60
Ca, %	<2.00	2.00–2.50	2.51–3.50	3.51–3.70	>3.70
Mg, %	<0.20	0.21–0.23	0.24–0.27	0.28–0.50	>0.50
Fe, ppm	<50	50–100	101–250	251–300	>300
Zn, ppm	<20	20–30	31–150	151–400	>450
Mn, ppm	<20	20–30	31–200	201–500	>500
Cu, ppm	<4	4–5	6–20	21–40	>40
B, ppm	<15	15–25	26–40	41–60	>60

[a] Compiled by Fregoni (6) using data from several authors.
[b] Values are on a dry weight basis.

being translocated from the leaves near the clusters into the berries. Water shortage and some injuries of roots or trunks caused mechanically or by parasites that hinder the absorption and transport of nutrients lead sooner or later to stunted growth and lighter colored leaves, which may be confused with N deficiency. Care should therefore be taken when prescribing corrective measures to address the real and not just the apparent cause of the disorder.

Nitrogen deficiency may be treated by supplying

Table 19.2. Average dry matter content of essential nutrient elements in grapevines[a]

Nutrient	Dry matter content
C, %	44.5
H, %	5.5
O, %	44.0
N, %	1.2
P, %	0.2
K, %	1.1
Ca, %	1.4
Mg, %	0.2
S, %	0.2
Fe, ppm	90.0
Zn, ppm	19.0
Mn, ppm	14.0
Cu, ppm	13.0
B, ppm	14.0
Cl, ppm	250.0
Mo, ppm	0.07

[a]Data are for White Riesling grapes on slate soils, 1,942 vines/acre (4,800 vines/ha); medium yield of 11,600 lb of grapes per acre (13,000 kg/ha).

Table 19.3. Essential nutrient elements taken from soil by grapevines[a]

Nutrient	Total removal per year			
	lb/acre	oz/acre	kg/ha	g/ha
C	...		...	
H	...		...	
O	...		...	
N	89		100	
P	16		18	
K	76		85	
Ca	103		115	
Mg	18		20	
S	13		15	
Fe		9.26		650
Zn		2.15		150
Mn		1.72		120
Cu		1.27		88
B		1.65		115
Cl		3.34		235
Mo		0.008		0.6

[a]Data are for White Riesling grapes on slate soils, 1,942 vines/acre (4,800 vines/ha); medium yield of 11,600 lb of grapes per acre (13,000 kg/ha).

nitrates to the vines by broadcasting and by foliar sprays, or on a long-term basis, use of ammonium-N sources including urea as well as organic substances such as manure.

Phosphorus (P)

Little is known about P deficiency of vines growing in the field. Most reports are based on symptoms obtained in nutrient culture free of P. The growth of shoots and roots is reduced, and leaves are small and dark green. Under severe deficiency, leaves turn reddish. The dormant buds contain small flower primordia that reduce the performance of the vines. On highly acid soils, a special type of P deficiency develops even in the presence of sufficient P (see Acidity Damage; Plate 288). Only a few experimental results report crop increases achieved through P fertilization. Because P is one of the major nutrients, it often is applied routinely at excessive rates in commercial operations. Hence, the soil in old viticultural regions frequently shows high P content. Excess P may induce Zn and Fe deficiency.

Potassium (K)

The symptoms of K deficiency vary with the stage of development of the leaf blades when K content in the tissue falls below the critical level. During the early part of the growing season the leaves lighten in color, with a few necrotic spots appearing along the margin of young blades. Dry weather induces necrotic speckles of varying form, number, and size, sporadically distributed in the interveinal tissue. Leaf margins dry and roll upwards or downwards, and blades become distorted and ruffled (Plate 283). Shiny round or elliptical spots appear on the upper side of mature leaves when the weather is sunny and dry. Such symptoms develop when the epidermis, now more light sensitive due to the K deficiency, loses its turgor under exposure to sunlight. At first the epidermis collapses without losing its color. A reflecting surface is formed under the cuticle, which appears as a shallow colorless lacquer or nail enamel stain. Neighboring spots unite and form larger light yellow shiny areas that turn brown within a few days (Plate 284). In late summer, the surfaces of older leaves at the base of shoots receiving direct sunlight become lilac-brown to dark brown ("black leaf") (Plate 285). The discoloration starts in the interveinal tissue and, in case of severe K deficiency, spreads over the entire leaf blade. The leaf margins of such brown leaves roll upward. On heavily cropped vines, the leaf browning is especially pronounced, because maturing berries become a K sink. Under deficiency conditions, K is translocated from the old leaves to the clusters. K deficiency retards the growth and ripening of clusters and berries (Fig. 19.1A). In late summer and fall, all symptoms may appear on individual leaves at the same time.

If vines suffer from K deficiency for several years,

their vigor decreases gradually. Finally, they produce only weak shoots with stunted, rolling, pale leaves. The shoots mature faulty and canes are easily killed below the freezing point. Grapevines with an insufficient supply of K are more susceptible to powdery mildew (*Uncinula necator* (Schwein.) Burrill). On sandy soils that have limited adsorption capacity and low pH values, an oversupply of K can induce Mg deficiency. This is frequently the case with young, noncropping vines, but it usually disappears as soon as vines start bearing fruit.

On soils with a large amount of K-fixing clay minerals (weathered micas, vermiculites, illites), it is recommended that the high fixation capacity be saturated before planting with a single massive rate of K up to 1,250 lb/acre K_2O (1,400 kg/ha) distributed uniformly in the future root zone. In this way, K deficiency can be avoided for years. The annual supply of K (chloride, sulfate, or nitrate) has to be selected according to the soil type and its K reserve, the grapevine cultivar, the rootstock, and the yield.

Calcium (Ca)

Calcium deficiency appears rarely on quartz, gravelly, or sandy soils that are strongly acid (pH < 4.5). A narrow, necrotic border first appears at the leaf margins and moves in steps toward the attachment point of the petiole and in extreme cases covers the entire leaf blade. On the primary bark of the shoot, dark brown pimples up to 0.04 in. (1-mm) in diameter may develop (Plate 286). Growth of the whole vine is affected by Ca deficiency. The growing clusters dry up starting from the tip, which is similar to severe stem necrosis.

Deep incorporation of dolomitic limestone in the root zone, preferably during winter, increases pH gradually and enriches the soil with Ca and Mg.

Magnesium (Mg)

Magnesium deficiency usually occurs on light, acid soils that have a low Mg content, although it is not limited to such situations. High applications of K or ammonium can induce Mg deficiency symptoms even on soils that contain enough Mg. Magnesium deficiency takes two forms. Before bloom, interveinal and marginal leaf necrosis dominate (Fig. 19.2A). During the summer and fall, interveinal yellowing or reddening and the development of brown-green patches are the major symptoms (Fig. 19.2B, Plate 287). As discoloration increases, it proceeds from the margin of the leaf in a wedge towards the the petiole attachment point. The interveinal yellowing has some similarity to the deficiency symptoms of Mn, B, and Zn. Magnesium surplus of grape vines in the field has no known typical visual changes.

The treatment of Mg deficiency rarely is absolutely indispensable. On acid soils, applications of dolomitic limestone are particularly useful. In general, magnesium sulfate (kieserite, Epsom salts) is more rapidly effective than magnesium carbonate. Foliar treatment requires three or four sprays after blossom time with a 2–5% magnesium sulfate solution.

Calcium and Magnesium

Two conditions can occur on acid soils when both Ca and Mg are deficient. They are acidity damage (*Säureschäden*) and stem necrosis (*Stiellähme*).

Acidity Damage. Leaf symptoms appear on vines on extremely acid soils (pH 3.4–4.5) with low Ca and Mg contents (Plate 288). Foliar analysis reveals simultaneous deficiencies of Ca, Mg, and P and an oversupply of K, Al, and especially Mn. Shortly after bloom, the margins of older leaves start to turn yellowish to light brown. Some of these spots connect with each other into larger, elongated rust-brown areas that have irregular outlines (Plate 288). Similar to Ca deficiency, the damaged area dies little by little. In dry weather, acid damage progresses rapidly. The clusters of

Fig. 19.1. Disorders of White Riesling cluster development; grapes grown under comparable soil and climate conditions were photographed on the same day. **A,** Potassium deficiency, seeded berries grow and ripen slowly. **B,** Boron deficiency produces small, seedless berries known as the "hen and chicken" or "pumpkin and pea" symptom or *millerandage*. Similar symptoms develop under B excess conditions. **C,** Normal White Riesling cluster as ripening begins.

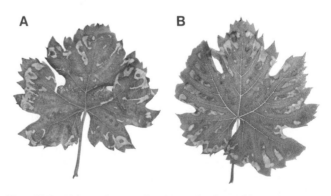

Fig. 19.2. Magnesium deficiency. **A,** Symptoms in spring: necrotic ring spots near the margin of the leaf blade; chlorotic and brown-green ring spots and straight patches along and between the veins. The interveinal areas are brightened; the leaf border remains green. **B,** Symptoms in summer: wedge-shaped necrotic ring spots and brown-green patches penetrate into the interveinal areas. (Also see Plate 287.)

grapevines with acid damage rarely mature fully. For treatment of acidity damage, apply dolomitic limestone to increase pH and to enrich the soil with Ca and Mg.

Stem Necrosis. Severe nonparasitic disturbances in the development of the clusters are termed stem necrosis. Shortly after the beginning of ripening when soluble solids range between 7 and 15% (30–60 Oechsle degrees), shallow or somewhat hollow (concave) elongated brown to black necrotic spots appear on the rachis and its branches (Plate 289). Less frequently, circular dark brown surrounding constrictions develop on the pedicels of the berries. Only if the necrosis reaches deeper tissue layers will distal parts of the rachis dry up. Stem necrosis is especially obvious in table grapes. During moist weather, the dead parts of the rachis may be infected by *Botrytis cinerea* Pers.:Fr., which then spreads to the healthy pedicels and berries. During the past 50 years, stem necrosis has increased substantially. The disorder is associated with a low Mg and Ca status. Spraying clusters with solutions of Ca and/or Mg (either chloride or sulfate) shortly before beginning of ripening is recommended as a preventive.

Sulfur (S)

Sulfur-containing organic and mineral fertilizers as well as fungicides usually add enough S to the soil to meet the need of grapevines. When applied as a fungicide for the control of powdery mildew under higher temperatures, S may produce serious burning on shoots, leaves, and clusters. Sulfur dioxide from industrial sources, deposited by wind onto moist leaves, can injure the leaf margins. The teeth on the margin of the leaves become necrotic similar to excess B symptoms (Plate 297). Exposure to sulfur dioxide during bloom may reduce fertilization (shot berries, poor set).

Iron (Fe)

Iron deficiency of green leaves produces yellowing and is usually called chlorosis, iron chlorosis, lime chlorosis, or lime-induced chlorosis. An actual lack of Fe is rare because most soils contain enough Fe to meet the requirements of grapevines. However, the low availability of Fe creates deficiencies. It is observed in all grape-growing regions of the world and it produces considerable economic loss. Its usual causes are unfavorable physical and chemical soil conditions.

Iron deficiency in sites with high-lime soils is obvious from the yellowing of the vines. The loss of chlorophyll starts between the small leaf veins. The blades show the most severe fading at their margins, and from there the yellowing enters the interveinal areas (Plate 290). The edges of severely yellowed leaves slowly dry, and the dead tissue rolls upwards. Finally, the leaves dry and fall off.

Lateral shoots develop with short, thin, pink-colored internodes. They have small, almost totally bleached, narrow, sharp-toothed leaflets that are folded along the central vein (Plate 291). With severe chlorosis, the delicate leaflets dry up, and eventually the shoot also desiccates starting at the tip. The reddish-colored shoots have thickened nodes that grow yellowed tendrils of almost normal size; these contrast sharply with the tiny deformed leaves. The flower clusters also become yellow. After bloom, poor set is frequently observed, and in extreme cases all berries are lost due to shatter (dropping of unpollinated flowers and young fruits from a grape cluster abut a week after bloom).

Iron deficiency is the most difficult of nutritional disorders to correct in plants. Soil improvement (loosening, aeration) is indispensable for a persistent causal treatment. Chlorosis-resistant vines and rootstocks are mostly very efficient. On calcareous soils, incorporation of high amounts of S may be tried. Iron chelates are often very effective but mostly limited in time. Repeated foliar sprays during the vegetation period with ferrous sulfate, citrate, or Fe chelates induce a regreening in spots.

Zinc (Zn)

Zinc deficiency exists in soils where Zn is absent or not available to the plant. At high P excess, Zn will combine with phosphate into the nearly water-insoluble Zn phosphate so that the roots cannot absorb it. Phosphate absorbed in excess leads to a partial inactivation of Zn, especially its specific function in the synthesis of auxins. The first symptoms of Zn deficiency can be observed before bloom on laterals and on shoot tips. The leaf blades are small, appear uplifted, and have wide open petiolar sinuses and sharp teeth. They are asymmetrical; one half of a leaf is always larger than the other. The main vein runs in a slight curve with the tip leading towards the small half. In the interveinal areas, a light mosaic pattern appears and continues to pale over time (Plate 292). Along the veins, a dark green border remains. With severe Zn deficiency, shoot growth length and caliper as well as maturation of shoots are affected. The yellow portions of the interveinal tissue die, whereupon the leaf blades roll in.

The intensity and distinctiveness of the symptoms vary somewhat between cultivars. A Zn deficiency causes high yield reductions, which in extreme cases lead to total crop loss. The berries remain smaller than normal; clusters are loose and have insufficiently lignified rachises (Fig. 19.3) that are likely to develop stem necrosis. Zinc sulfate solutions (up to 5%) supplied to the roots or as sprays (0.5%) to the leaves can rapidly cure Zn deficiency.

Manganese (Mn)

Manganese deficiency is observed mainly on alkaline, sandy soils containing high amounts of humus and on Mn-poor limed soils. In early summer, the leaves at the base of the shoot start to pale, and shortly afterwards small, polygonal yellow spots appear in the interveinal tissue. These are arranged in a mosaic pattern bordered by small green colored veins. Only a very small seam along first- and second-order veins remains green (Plate 293). Near fall the discolored leaves turn bronze. During drought the margins of older leaves dry up while the younger leaves at the tips of shoots and on laterals remain green. The symptoms are more severe on sun-exposed leaves than on shaded leaves. The growth of shoots, leaves, and berries is affected by Mn deficiency, and the maturation of the clusters is delayed. The symptom of Mn deficiency on soils high in lime is frequently concealed through the more severe yellowing from lime chlorosis (Fe deficiency) occurring simultaneously. Foliar spray with solutions containing manganese chloride or sulfate can prevent and cure Mn deficiency. On highly acid soils, Mn can be in excess (see Acidity Damage).

Copper (Cu)

The symptoms of Cu deficiency have been described in Australia as weak root growth, narrow pale green leaves, short shoot internodes, rough bark, and small crops of low quality.

Boron (B)

Boron deficiency is one of the most serious nonparasitic grapevine diseases. It is observed especially on strongly acid soils (pH 3.5–4.5). The first symptoms appear before bloom on tendrils near the shoot tip and on stalks of inflorescences. Dark, knotty bulges form and become necrotic. The distal portions dry up and flower clusters die. During rapid shoot growth, younger internodes swell slightly and darken at one or several places (Plate 294), and the pith becomes necrotic (Fig. 19.4). Usually the part of the shoot that is distal to the swelling dies. Older

internodes develop swellings with deep folds and pits that often reach the pith (Fig. 19.5). These swellings are the most characteristic B deficiency symptoms. Deformed internodes are thicker and shorter than normal ones. During long periods of drought, several short internodes will follow one after another (Fig. 19.6). The leaves on such nodes have short, thick petioles, sometimes showing longitudinal tears or necrotic pits.

Severe B deficiency in summer and fall may lead to apical necrosis of the shoot primordium in the dormant buds. In the following season, such damaged buds produce short, bushy, branched sterile shoots.

Boron deficiency symptoms on the leaves appear first as lighter areas in a mosaic pattern, and the fine leaf veins in the interveinal areas turn brown. On the margin and between the veins, red-brown spots appear and slowly darken (Plate 295). The main veins near them usually remain green. After long periods of B deficiency, these veins break open lengthwise on both sides of the leaf.

Boron deficiency affects the development of berries and clusters (Fig. 19.1B). Only a few seeded berries set, and small, seedless berries develop (called *millerandage,* hen and chicken, or the pumpkin and pea symptom). If acute B deficiency starts after bloom, the berries develop necrotic foci inside, which account for their gray-

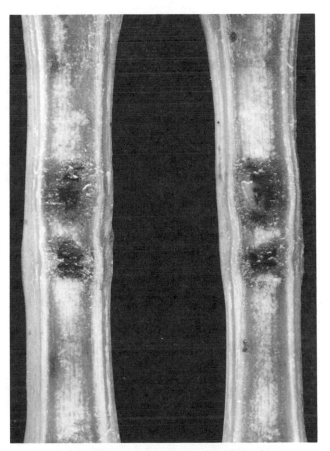

Fig. 19.4. Boron deficiency manifested in longitudinal sections of two nodelike intumescences showing necrotic foci in the medullary tissue.

Fig. 19.3. Zinc deficiency causes a loose (straggly) cluster with seeded berries of varied sizes and a weak stalk.

brown color. This symptom may be confused with the late infection of *Plasmopara viticola* (Berk. & M.A. Curtis) Berl. & De Toni in Sacc. (brown rot).

Boron deficiency also affects the growth of roots, which remain short, thicken, swell up into knots, and break open longitudinally (Plate 296).

Boron deficiency is easy to cure by supplying B-containing fertilizers (borax, sodium borate, borated superphosphate). Boric acid and sodium borate are suitable for foliar sprays.

Molybdenum (Mo)

Molybdenum deficiency in the field has not been observed to date. In nutrition experiments, yellowing and necrosis of leaf blades occur in the absence of Mo.

Chlorine (Cl)

Nothing is known about Cl deficiency of grapes.

Toxicity Symptoms

Nitrogen

Grapevines tolerate high N so long as other macro- and micronutrients are in adequate supply. An excess of N increases growth, the internodes become long and

of large caliper, and leaf blades turn deep green, thick, and sometimes crinkled. Shoot growth may be prolonged, causing poor maturation and increased cold susceptibility. The tissue of all organs is soft and spongy and provides little resistence to the hyphae of certain fungi, e.g., *Botrytis cinerea*.

Boron

An oversupply of B affects the development of all aboveground parts of the vine. The leaves are severely malformed and necrosis develops in the interveinal areas and also on the turned down, toothless margins of older leaves (Plate 297). The tip growth of the main shoots decreases in favor of the laterals, which produces vines that look frail and bushy. Berry and cluster growth are affected similar to that of B deficiency.

Chlorine

Chloride excess produces injury on salty soils or if vines are irrigated with high chloride water. The leaf margins roll downward and die with red-brown discoloration. The marginal necrosis proceeds toward the center of the leaf blade and sometimes covers it totally. Dead leaves fall off, stimulating the development of lateral shoots and sometimes the emergence of shoots from dormant buds. The wood does not mature sufficiently,

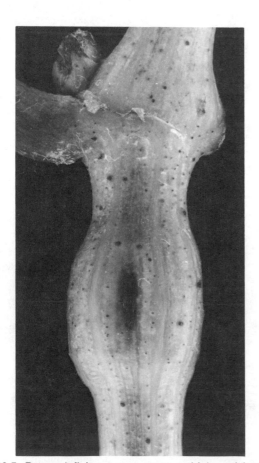

Fig. 19.5. Boron deficiency may cause a thickened internode with a deep pit reaching the pith.

Fig. 19.6. Extremely stunted internodes and necrotic apex in a B-deficient plant. Leaves emerged from the affected zone are small and deformed; buds are inclined to leaf out prematurely.

which leads to problems with shoot emergence the following season. In extreme cases, Cl excess may cause the death of the vine. Old vines are more sensitive to Cl than young vines. Spraying grapes with excessive concentrations of calcium chloride and/or magnesium chloride solutions for the prevention of stem necrosis can cause burning of the leaf border (Plate 298).

Copper

Where Cu compounds were or still are used as fungicides, the top 8 in. (20 cm) of soil is enriched with Cu. Since most vine roots are found deeper, where Cu content is lower, no damage has been observed in commercial plantings. However, in grapevine nurseries, considerable disorder in the growth of roots and eventual death of roots occur if Cu has been used over many years as a fungicide. Sprays containing Cu may cause considerable burning on shoots, leaves, and flower clusters; hence, check with a viticulturist for rates and conditions to use. Copper deposited on the stigma will prevent the germination of pollen and thus inhibit fertilization. With hermaphrodite cultivars, this effect leads to the development of shot berries and poor set.

Molybdenum

Molybdenum excess leads to a chlorosis that is difficult to differentiate visually from the symptoms of Fe deficiency.

Others

Toxicity symptoms that cannot be easily identified will often appear on plants. Toxic emissions from industrial plants are a possible cause. Plate 299 shows the effect of hydrogen fluoride emissions.

Selected References

1. Bovey, R., Gärtel, W., Hewitt, W. B., Martelli, G. P., and Vuittenez, A. 1980. Virus and Virus-like Diseases of Grapevines: Colour Atlas of Symptoms. Editions Payot, Lausanne, Switzerland.
2. Champagnol, F. 1984. Éléments de Physiologie de la Vigne et de Viticulture Générale. Champagnol, Saint-Gely-du-Fesc, France.
3. Christ, E. G., and Ulrich, A. 1954. Grape nutrition. Pages 295-343 in: Fruit Nutrition. Horticultural Publications, Rutgers University, New Brunswick, NJ.
4. Christiansen, L. P., Kasimatis, A. N., and Jensen F. L. 1978. Grapevine nutrition and fertilization in the San Joaquin Valley. Div. Agric. Sci. Univ. Calif. Publ. 4087.
5. Cook, J. A. 1966. Grape nutrition. Pages 777-812 in: Nutrition of Fruit Crops. Childres, N. F., ed. Horticultural Publications, Rutgers University, New Brunswick, NJ.
6. Fregoni, M. 1980. Nutrizione e Fertilizzazione della Vite. Edagricole, Bologna, Italy.
7. Gärtel, W. 1974. Die Mikronährstoffe—ihre Bedeutung für die Rebenernährung unter besonderer Berücksichtigung der Mangel—und Überschusserscheinungen. Weinberg Keller 21:435-508.
8. Smith, C. R., Shaulis, N., and Cook, J. A. 1964. Nutrient deficiencies in small fruits and grapes. Pages 327-357 in: Hunger Signs in Crops: A Symposium. H. B. Sprague, ed. David McKay, New York.

Part V

Turfgrass

Chapter 20

Thomas R. Turner
Department of Agronomy
University of Maryland, College Park

Turfgrass

While nutritional problems of turfgrasses are quite common, extreme deficiencies manifested by classical leaf symptoms are much less common. Much more common and nonacute symptoms of turfgrass nutritional problems are reductions in shoot growth, root growth, rhizome or stolon production, stand density, and visual quality as affected by wear stress, environmental stresses, and incidence of pest problems. Thus, it is important to recognize some of the typical growth and quality effects of nutritional deficiencies, since classical leaf symptoms are the exception rather than the rule.

Although it is a practical indicator in virtually all other agronomic crops, economic yield is not a good guideline for determining the degree of a nutrient deficiency in turfgrasses. The primary goals of successful turfgrass management are qualitative rather than quantitative. High yield can actually be detrimental rather than a desired goal. Thus, in assessing nutritional deficiencies of turfgrass, a variety of aesthetic and functional characteristics are much more important than top growth rates. For example, low levels of a nutrient may result in the increased incidence of a turfgrass disease, which in turn reduces turfgrass quality. While this may not meet classical growth guidelines for a nutrient deficiency, the practical effect of a low nutrient level enhancing the incidence of a disease qualifies it as a nutrient deficiency in turfgrass.

Many factors complicate the degree of nutrient deficiencies and when they may occur in turfgrasses. These factors include turfgrass species and cultivar; length of growing season; inherent soil nutrient levels, soil texture, and soil organic matter levels; degree of environmental, pest, and wear stresses; amount of rainfall and irrigation; management practices such as clipping removal versus recycling; and quality expectations at a specific site.

Deficiency and Toxicity Symptoms

Nitrogen (N)

Nitrogen deficiency in turfgrass leaves is exhibited by a general yellowing or chlorosis, initially in the older leaves (Plate 300). Tip dieback and decreases in shoot density and tillering follow. Low N is undoubtedly the most common nutritional deficiency of turfgrasses encountered. This fact is due to the relatively high N requirements of most turfgrass species, the inability of most soils to meet these requirements without fertilizer N applications, and the dramatic impact N has on many of the most important components of turfgrass quality. As a result, more research has been conducted on turfgrass response to N applications than on all other nutrients combined, with much of the information now categorized as common knowledge. Total N requirements to produce maximum turfgrass quality vary with both species and cultivar (28,32).

The most universal effect of N applications to turfgrass is an increase in shoot growth rate and enhancement of a darker green. Increases in N levels above zero also can result in increases in root growth (11,18,44,54) and rhizome production (41,50). However, excess available N can cause excessive shoot growth and a subsequent reduction in root growth (11,17,54,78,89,90). Nonstructural carbohydrates can also be reduced by an excess of available N (78,89,90).

Increases or decreases in shoot production, root production, and carbohydrate levels can directly and indirectly affect many other factors influencing turfgrass quality. Turfgrass wear tolerance may be improved by N fertilization up to a threshold point, beyond which additional N will reduce wear tolerance (11,59). Turfgrass recovery from wear injury may be improved by N fertilization (41,57) but may also be inhibited by excess N (57). Recovery from other stresses, such as scalping (90), herbicide injury (47), drought (79), and winter dormancy (16) may also be enhanced by N applications.

Nitrogen fertility can have an important effect on the tolerance of turfgrasses to environmental stresses. Reductions in heat tolerance (74), drought tolerance (90), and cold hardiness (12) have been related to excessive N.

Many turfgrass pest problems may be associated with N fertility. The increase in turfgrass growth rates and density caused by N fertilization generally results in reduced weed encroachment (1,70,81). Excess N levels can lead to reductions in turfgrass density, which result in increased susceptibility to weed encroachment. Many turfgrass diseases are more prevalent under low N fertility. These include dollar spot (*Sclerotinia homoeocarpa* F. T. Bennett) (20,29,62), red thread (*Laetisaria fuciformis* (McAlpine) Burdsall) (10,38,84,93); rusts (*Puccinia* spp.) (21), take-all patch (*Gaeumannomyces graminis* (Sacc.) Arx & D. Oliver) (25), and yellow tuft (*Sclerophthora macrospora* (Sacc.) Thirumalachar, C. G. Shaw & Narasimhan) (26). High N levels versus moderate N levels, however, may increase turfgrass susceptibility to several other diseases, such as Rhizoctonia brown patch (*Rhizoctonia solani* Kühn) (6), anthracnose (*Colletotrichum graminicola* (Ces.) G. W. Wils.) (22), and Drechslera leaf spot (*Drechslera sorokiniana* (Sacc.) Subramanian & P. C. Jain) (14,61).

Phosphorus (P)

The most common leaf symptom of P deficiency is a dark green color that progresses to a purplish to reddish purple color, particularly on the older leaves. The overall turfgrass stand often appears wilted, and the symptoms are sometimes confused with the onset of drought stress. Excess P may result in chlorotic turf (82), reduced turfgrass quality and chlorophyll content (17), and reduced top and/or root growth (5,42,66).

Although the leaf symptoms associated with extreme P deficiency may not be exhibited, turfgrass quality and performance can be seriously affected by inadequate soil P. Perhaps the most common occurrence of inadequate soil P affecting turfgrass is during the establishment phase.

For warm-season grasses, adequate P has been reported to be necessary for improvement in the rate of vegetative establishment of zoysiagrass and St. Augustinegrass (50,92). On soil with high initial soil P levels, however, establishment rate benefited little from P fertilization of Meyer zoysiagrass plugs (31).

Phosphorus is critical to the establishment of cool-season grasses from seed. Insufficient P during seedling establishment is one of the most common nutrient deficiencies in turfgrasses (Plate 301). Several researchers reported improvements in seedling growth and/or density from applications of P to the seedbed (57,64,91).

Turner (84) initiated an extensive series of soil test calibration studies for establishment of cool-season grasses. A review of these studies indicated that soil P levels of 60 lb/acre (67 kg/ha) were not sufficient for optimum establishment rates. Differences in turfgrass stand density began to become apparent 3 to 4 weeks after germination and were extremely evident after 6 weeks. Presumably, differences did not occur in the first 3 weeks because seed P was utilized. Phosphorus applications of 115–356 lb/acre (130–400 kg/ha) to the seedbed, resulting in soil P levels of 72–140 lb/acre (80–157 kg/ha), provided the most rapid establishment.

Although significant P deficiencies, as manifested by turfgrass density and establishment rates, were obtained in these studies, the classical leaf symptoms of P deficiency were not observed.

Turfgrass species and cultivars have been shown to respond differently to seedbed P applications and thus have different P requirements. It has been found (85) that the magnitude and duration of responses to seedbed applications of P to the same soil were different for Chewings fescue, perennial ryegrass, and Kentucky bluegrass. Soil P levels were considered less important for Chewings fescue. At a soil pH of 4.5, Kentucky bluegrass exhibits positive growth responses to seedbed P applications, whereas there is little influence on red fescue (51). Neither species was influenced by seedbed P applications at a soil pH of 6.5. Merion Kentucky bluegrass was more sensitive to seedbed applications of P than common Kentucky bluegrass.

The long-term effects of seedbed P applications on soil P levels and turfgrass performance were also investigated (86). Six years after seedbed P applications and with no further P applications, spring greening of Kentucky bluegrass was enhanced, and crabgrass encroachment was reduced on plots that had received seedbed P applications. Greater top growth and reduced dandelion encroachment were reported 9 years after these seedbed applications of P had been made. In another study (84), much higher crabgrass populations were found one year after turfgrass establishment in plots receiving no seedbed P, presumably due to reduced turfgrass density and thus competition against crabgrass.

Phosphorus deficiency on established turfgrass stands is less common than during the establishment phase. Probably the most frequent occurrences of moderate to severe P deficiency on established turf are on very sandy soils, such as putting greens that have been highly modified and have a high sand content (80% or more) (Plates 302 and 303).

Although some studies have shown applications of P increased established turfgrass top and/or root growth (39,46), most studies have shown little response even on soils with low available P (19,75,84,87).

Factors affecting quality of turfgrass have been more commonly observed to be affected by P applications. Both cool- and warm-season turfgrasses fertilized with P where soil P levels are low are generally a lighter green, although healthier (82,84,87). The apparent darker green color of P-deficient turfgrass is due to the reddish-purplish color of older leaves, which becomes objectionable as the deficiency becomes more severe and affects more leaves. On soils low in P, however, little influence on general turfgrass quality was found for perennial ryegrass, Kentucky bluegrass, or creeping red fescue (84) or creeping bentgrass (19).

Heat, drought, and cold tolerance do not appear to be affected greatly by P levels alone (73,84). However, interactions with other nutrients may have some importance. If N and K were in adequate supply, P applications somewhat improved the cold tolerance of bermudagrass (33). Adequate P has been shown (79) to be essential for recovery of Kentucky bluegrass from drought stress, particularly when higher N applications had been made.

Some diseases that can have major impacts on turfgrass quality have been shown to be influenced by P applications, although these effects generally appear minor compared with those of some other nutrients, particularly N and K. Kentucky bluegrass fertilized with N and P, N and K, or most dramatically N, P, and K had a lower incidence of stripe smut than turf fertilized with N alone (45). Phosphorus applications may interact with other nutrients in reducing the incidence of Fusarium patch (37), take-all patch (23), and red thread (34), although others have found no influence of P on red thread incidence in perennial ryegrass (10) or creeping red fescue (84).

Maintenance applications of P may influence the relative proportion of species in a turfgrass mixture, particularly annual bluegrass and creeping bentgrass mixtures. It has been reported (35,95) that the amount of annual bluegrass in creeping bentgrass turf increases as soil P levels are raised by P fertilization. Phosphorus applications have been found to increase both the density and survival of annual bluegrass maintained under putting green conditions. Phosphorus levels needed to be higher under higher N levels to maximize survival of annual bluegrass. The drought resistance of annual bluegrass, which is very poor compared with bentgrass and most other species, was enhanced by P additions (30). These studies suggest that the P requirements of annual bluegrass are higher than bentgrass. Additionally, annual bluegrass produces much viable seed even under putting green conditions. Higher P levels may increase the proportion of subsequent annual bluegrass seedlings that survive and mature.

Potassium (K)

Turfgrass leaf symptoms of K deficiency generally include yellowing of older leaves followed by tip dieback and necrosis along the leaf margin. Early spring chlorosis in such a manner is often indicative of K deficiency (Plate 304). Excess K can result in a burning of turf from the creation of high soluble salt levels (84). Theoretically, excess K could also induce Ca or Mg deficiencies, although this effect has rarely been documented in turfgrasses. It has been reported (17) that under higher N, tissue growth of bentgrass decreased somewhat with added K. It was postulated that a Mn deficiency induced by a combination of N and K may have been the cause.

Unlike P, application of K has shown few benefits for turfgrass establishment, and deficiencies have rarely been reported (31,84) or observed. Juska (50) found that Meyer zoysiagrass top growth and stolon growth and spread during establishment were enhanced by K applications.

Growth responses of established turfgrasses to K applications have also not been dramatic. Turner (84) reported that the top growth of creeping red fescue and Kentucky bluegrass grown on soils low in K was not affected by K applications, whereas perennial ryegrass top growth was affected minimally and on only 33% of the dates measured. Moderate bentgrass top growth responses to K applications were reported (87) on a low-K soil for the initial 4 years of a study, but none from the sixth year on compared with plots receiving no K (19). One study reported (19) that more K was needed to maximize the quality of Kentucky bluegrass and bentgrass than was needed to maximize growth, while others (75) found that the growth response of warm-season turfgrasses was species dependent, with centipedegrass having higher K requirements than bermudagrass or zoysiagrass. Keisling (56) found that on a low-K soil, the initiation of new bermudagrass rhizomes and the longevity of existing rhizomes were directly related to K applications.

Whereas severe K deficiency symptoms have also rarely been reported for established turfgrasses, many beneficial responses have been observed as a result of K applications. These responses indicate that mild deficiencies are rather common.

Poorer color and early spring performance of both zoysiagrass and bermudagrass were reported (82) when no K was applied. Similar effects were noted on creeping bentgrass (87). Early spring chlorosis lasted 10 to 14 days on plots receiving no K on soil very low in K.

Potassium appears to play an important role in tolerance of heat, drought, and cold of turfgrasses. Whereas no leaf symptoms of K deficiency may be apparent, beneficial responses in stress tolerance and thus turfgrass quality may result from K applications. Numerous researchers have reported decreases in winter injury and/or increases in winter hardiness of centipedegrass (73) and bermudagrass (1,3,33,54) after K applications. It was suggested (4) that winter survival of turfgrass was at a maximum when applied K rates were approximately half that of applied N rates.

Bentgrass wilting has been shown to be more severe during drought and heat stress on a low-K soil where no K was applied. Recovery of Kentucky bluegrass from summer drought was enhanced by K applications, particularly when coupled with P fertilization (79). Under high N fertility it has been reported (74) that higher K rates enhanced heat resistance of Kentucky bluegrass. On plots receiving no K, the first indication of a K deficiency did not occur until after 20 years, at which point a decrease in drought resistance was observed.

Calcium (Ca)

Signs of Ca deficiencies occur first in younger turfgrass leaves, which exhibit a reddish brown color along leaf margins. Few studies have been made of turfgrass performance under Ca-deficient conditions. However, differences among Kentucky bluegrass cultivars have been found in tolerance to low Ca levels (71). Also, Pythium blight incidence of colonial bentgrass was shown to be greater in a Ca-deficient nutrient solution culture (69).

Actual Ca deficiencies in turfgrass are probably rather rare, and problems associated with low soil Ca may usually be attributed to soil reaction problems rather than a true Ca deficiency. These problems may be due to direct hydrogen ion toxicity; increased availability of nonessential elements in toxic amounts; increased or decreased availability of essential nutrients to toxic or deficient levels; and reductions in soil microbial populations and their subsequent effect on disease incidence, N availability, and thatch decomposition. Thus, the use of a liming agent to reduce soil acidity, not the addition of Ca alone, is needed to enhance turfgrass performance. For example, the application of gypsum ($CaSO_4$), which adds Ca to the soil but has little impact on soil reaction, is of little benefit in improving turfgrass performance on acid soils.

Magnesium (Mg)

Signs of Mg deficiencies occur in older turfgrass leaves first, with leaves turning red to cherry red along margins. Magnesium deficiencies in turfgrass are rare, and little research has been undertaken relating soil Mg to turfgrass performance.

Among warm-season turfgrass species, Kamon (55) reported that Mg deficiency symptoms were greatest for zoysiagrass, with centipedegrass intermediate, and bermudagrass least. The onset of Mg deficiency symptoms for these three grasses, however, were in the opposite order. Sartain and Dudeck (77) measured no significant growth response of bermudagrass or perennial ryegrass to Mg applications of 120 lb/acre (134 kg/ha) on soil. Beneficial response was not found from Mg applications to Kentucky bluegrass growing on a low-Mg soil (D. V. Waddington, Pennsylvania State University, *personal communication*).

Sulfur (S)

Sulfur deficiencies in turfgrass leaves occur in older leaves first and are very similar in appearance to N deficiency. In some cases, however, the midvein may remain green when S is deficient in contrast to an N deficiency.

Whereas severe S deficiencies exhibiting leaf symptoms previously described are rare, beneficial turfgrass responses to S applications have been reported in several studies. Compared to no S applications, a 71% increase in bentgrass growth was measured (36) when S applications were made to bentgrass under high N fertility. Improvements in bentgrass color were observed (7) when S applications were made, and enhancement of bentgrass color from N applications was appreciably better when coupled with S applications.

Although not a true deficiency phenomenon, several factors that influence turfgrass quality may be affected by S applications. Incidence of Fusarium patch (7) and take-all patch (23) of bentgrass have been reported to be reduced by S applications. Dernoeden (25), however, did not find that S alone was effective in the control of take-all patch of bentgrass. It was also found (35) that, depending on N fertility levels, S applications could reduce annual bluegrass populations in bentgrass as well as prevent black algae encroachment.

Iron (Fe)

Iron deficiency in turfgrass leaves is first manifested in younger leaves with typical symptoms including an interveinal chlorosis (Plates 305 and 306). Under severe deficiencies, leaves may be almost white. As a result of reduced turfgrass vigor, Fe-deficient plants may lack an upright growth habit, which results in a matted stand that may be difficult to properly mow (67).

Conditions which may induce or lead to Fe deficiency problems include high soil pH (58,76), high levels of soil P (17), high rates of N fertilization or other factors leading to rapid growth rates, excess levels of other micronutrients, sandy soils, and cold, wet soils.

Differences among turfgrass species and cultivars have been found in relation to potential Fe deficiency problems (40,58,63). These differences offer a means of minimizing the likelihood of Fe deficiencies through the careful selection of species and cultivars, particularly for use in situations where deficiencies are a common problem. These known differences also suggest that further breeding and selection work may be warranted in this area.

Typically, Fe deficiencies have been corrected through the application of various Fe-containing compounds, although reducing soil pH may be sufficient on higher pH soils (43). Color enhancement from foliar applications of Fe may occur within hours. The dramatic increase in green color from the application of low rates of Fe to Kentucky bluegrass was found (24) to last 4–5 weeks, with slight improvements lasting up to 16 weeks. Minner and Butler (67) observed improvement in Kentucky bluegrass at rates up to 44 lb/acre (49 kg/ha). The duration of color enhancement can be related to weather. Color enhancement due to Fe applications was found (95) to last only 2–3 weeks during cool, wet periods, whereas it lasted several months during cool, dry periods.

Effects other than color enhancement have also been reported. Although clipping weights are generally not greatly affected by Fe applications (24), increases in root growth of Kentucky bluegrass were reported (24) with applications of 1.0 lb/acre (1.1 kg/ha) of Fe. Iron applied with N, compared with N alone, improved appearance, chlorophyll content, and early spring growth of creeping bentgrass (81). In addition, recovery from winter desiccation was improved, and the injurious effects of heavy fall-winter N applications were offset.

Toxicity problems with excess Fe applications or Fe applications under stressful environmental conditions have been observed frequently and reported by several researchers. Toxicity symptoms usually are a blackening of leaf tissue (Plate 307) without permanent damage to the turfgrass plant (15,95). It was found (24), however, that the application of 50 lb/acre (56 kg/ha) of Fe to Kentucky bluegrass caused severe discoloration, burning, and inhibition of rhizome formation. Sod density was adversely affected for 10 weeks. Foliar dieback has been observed (95) although not death of plants, at Fe applications rates in excess of 16.7 lb/acre (17.7 kg/ha). The degree of toxicity on centipedegrass for a particular rate of Fe has been found (13) to be dependent on the air temperature on the day of application and on N levels.

Zinc (Zn)

Turfgrass plants exhibiting Zn deficiency may have stunted leaves, puckered leaf margins, and some chlorosis. As with many of the micronutrients, Zn effects on turfgrass have not been extensively studied.

Beneficial effects were not observed from Zn applications to Penncross creeping bentgrass grown on soil mixtures containing 80% sand (17; N. E. Christians, Iowa State University, Ames, *personal communication*). Other researchers (24) did not find any effect from Zn applications to Kentucky bluegrass in regard to growth rate, color, and density. However, they did report that root growth was stimulated by low rates of Zn, whereas rhizome growth was strongly inhibited by a high application rate of 25 lb/acre (28 kg/ha). N. E. Christians (*personal communication*) did not observe detrimental effects to bentgrass from Zn applications as high as 47–63 lb/acre (53–71 kg/ha). Differences in bermudagrass tolerance to excessive Zn levels in solution culture have been reported (94).

Manganese (Mn)

Interveinal chlorosis of younger turfgrass leaves develops with a Mn deficiency. Necrotic spots may develop. Although not uncommon on other crops, Mn deficiencies have rarely been observed on turfgrass.

On bentgrass, tissue growth was reported to decrease somewhat under higher N fertility when combined with K additions (17). A Mn deficiency induced by the N and K combination was hypothesized as the cause. Tissue Mn levels of 33 ppm were measured, which were at the low end of the sufficiency range suggested by Jones (49). Under low fertility conditions, low rates of Mn were found (24) to improve the root growth and color of Kentucky bluegrass for about 4 weeks. On both calcareous and acidic sands, it has been found (N. W. Hummel, Cornell University, *personal communication*) that repeated small applications of Mn to creeping bent-

grass improved growth and quality, although benefits were greater on the acidic sand. Effects were dependent on the N level, with detrimental effects being found from high Mn applications when N was low. No effects were found (84) from applications of Mn at varying rates to creeping bentgrass grown on a sand-peat mixture.

Copper (Cu)

Although not well defined, turfgrass leaf symptoms of Cu deficiency include a tip dieback of younger leaves, which may become white. Deficiencies in turfgrass appear to be rare. Little research has been performed relating Cu levels to turfgrass performance. Beneficial responses were found from Cu applications to some grasses grown on Histosols in Florida in the 1920s (80). Toxicities also appear to be rare, although it was found (94) that with solution concentrations above 0.25 ppm of Cu, root injury of bermudagrass occurred. Different tolerances of bermudagrass cultivars to excess Cu were also reported.

Boron (B)

Individual leaf symptoms of B deficiency have not been well defined. Reduced growth and stunting of plants are symptoms commonly associated with B deficiency.

Little research has been conducted evaluating the effects of B on turfgrass. It was found (24) that where fertility levels were low, B applications caused a rapid greening in Kentucky bluegrass that lasted for 5 weeks. Root growth and sod density were also enhanced for about 5 weeks. However, application of 7.5 lb/acre (8.4 kg/ha) of B in conjunction with high fertility caused a pale green color for 13 weeks as well as leaf tip death. Excessive levels of B have been reported to produce a tipburn (72), with injury being greatest in creeping bentgrass, followed by perennial ryegrass, tall fescue, Kentucky bluegrass, zoysiagrass, and bermudagrass. Clipping will remove excess B that accumulates in leaf tips. Thus, increasing mowing frequency may help alleviate B toxicity problems.

Molybdenum (Mo)

Molybdenum deficiency symptoms are typically similar to those of N deficiency, although some interveinal chlorosis may occur. The author is unaware of any published research investigating Mo deficiencies or toxicities in turfgrass.

Critical and Sufficient Tissue Concentration Levels

Nutrient levels in turfgrass vary considerably depending on species, cultivar, time of sampling, and management practices. General sufficiency ranges for turfgrass are listed in Table 20.1 (50). However, to show the variation that can be experienced, Table 20.2 provides values for several nutrients for various turfgrass species and cultivars as reported by two researchers (9,88).

Nitrogen

Wide ranges of tissue N may be measured in turfgrass plants depending on turfgrass species, cultivar, time of sampling, and management practices. For example, tissue N has been found to be inversely proportional to the mowing height and directly proportional to N fertilization rate (83). Tissue N levels were reported to be affected by K fertility as well (68). A 2.94–4.24% range of tissue N reported in Kentucky bluegrass (39) depended on the time of sampling during the growing season. Others (9) reported a range of tissue N among turfgrass species and cultivars grown under the same conditions from 2.0 to 6.0%. Jones (49) suggested a general turfgrass sufficiency range for tissue N of 2.75–3.5%. The actual ideal range may be dependent not only on the species but also on the turfgrass quality level desired, whether a site will be irrigated, and the intensity of use of the site.

Phosphorus

Wide ranges of P have been reported in the tissue of turfgrasses. Tissue P levels can vary substantially with species and cultivar. Under identical management, tissue P levels as low as 0.2% for Meyer zoysiagrass to as high as 0.51% for Kentucky-31 tall fescue have been reported (9). Bluegrass cultivars in this same study differed by 0.18%. Additional P levels are listed in Table 20.2.

Variation in tissue P levels throughout the season can also make interpretation of these levels difficult. The average tissue P levels for 15 Kentucky bluegrass cultivars were found to range from 0.10 to 0.44% over a 1-year sampling period (65). In clipping yields of Merion Kentucky bluegrass samples taken in July and September, 74% of the samples having less than 0.36% tissue P were related to yields less than 50% maximum (39). However, when samples were taken in November, only 60% of the samples having less than 0.23% P were related to yields less than 50% maximum. Thus, sampling time becomes an important factor in determining whether a given tissue P level is in a deficient or sufficient range.

Potassium

As with P, levels of tissue K can vary dramatically depending on turfgrass species, time of sampling, and management practices. Tissue K levels for various species are given in Table 20.2. All received the same management. In 15 Kentucky bluegrass cultivars, tissue K varied from a low of 2.08% for Parade and Sydsport to a high of 2.57% for Enmundi (65). In this same study, sampling time during the year had an even more dramatic effect on tissue K levels. Tissue K for Enmundi varied from 1.76 to 3.35% in a single season.

Relating K deficiency symptoms to tissue K levels has not been widely documented or particularly successful. No deficiency symptoms were observed (79) on perennial ryegrass, Kentucky bluegrass, or creeping red fescue with respective tissue K levels of 2.9, 1.9, and 1.8%. Early spring chlorosis of putting green creeping bentgrass was associated with tissue K levels of 5.8% (91), whereas no chlorosis was associated with tissue K levels reaching 10.4%. A tissue K sufficiency range of 1.0–2.5% has been suggested by Jones (49), although he stressed that these ranges were not universally applicable.

Calcium

Tissue levels of Ca tend to be similar in magnitude to tissue P. As with most elements, tissue Ca levels vary considerably among turfgrass species and with time of year. Reports of tissue Ca levels are given in Table 20.2. Tissue Ca levels of Kentucky bluegrass were reported (39) to range from 0.82 to 1.47% during one growing season.

Little research information has been generated that relates Ca deficiencies in turfgrasses to specific tissue levels. Sufficiency levels for turfgrass tissue Ca in the range of 0.5–1.25% have been suggested (49).

Magnesium

Although tissue Mg levels tend to be lower in turfgrasses than tissue Ca or P, variation among species is of a similar relative magnitude. As shown in Table 20.2, tissue levels of Mg vary more than twofold. A large variation was found for tissue Mg in Kentucky bluegrass with values ranging from 0.24 to 0.46% during one growing season (39). Although little direct research information is available for turfgrasses, a general turfgrass sufficiency range for tissue Mg of 0.2–0.6% has been suggested (49).

Sulfur

In contrast to those of other macronutrients, tissue S levels in turfgrasses have not been commonly reported. Tissue S levels of 0.15% were reported to be sufficient

Table 20.1. General sufficiency range for turfgrass[a]

Nutrient	Range
N, %	2.8–3.5
P, %	0.1–0.4
K, %	1.0–2.5
Ca, %	0.5–1.2
Mg, %	0.2–0.6
S, %	0.2–0.4
Fe, ppm	35–100
Zn, ppm	22–30
Mn, ppm	25–150
Cu, ppm	5–20
B, ppm	10–60

[a] Data from Jones (50).

for creeping bentgrass and creeping red fescue (60). A range of 0.30–0.42% S was found in creeping bentgrass (36), with the level being dependent on the rates of N, P, K, and S fertilization. Depending on the time of sampling during the year, it was found that tissue S levels in creeping bentgrass range from 0.26 to 0.4% (87). Jones (49) suggested a turfgrass sufficiency level for tissue S of 0.2–0.45%.

Iron

Turfgrass species grown under identical conditions exhibit a very wide range of tissue Fe levels (Table 20.2), suggesting that interpretation of tissue Fe levels for diagnostic purposes could be very difficult and certainly must take into account species differences.

A tissue Fe sufficiency range of 35–100 ppm for turfgrasses has been recommended (49). Brown (8) suggested that general tissue ranges for Fe-deficient, normal, and Fe-toxic plants were 10–30, 60–300, and 400–1,000 ppm. A minimum acceptable level of 50 ppm of Fe has been suggested (9).

Zinc

Tissue levels of Zn in turfgrasses tend to be lower than those of Fe (Table 20.2). No deficiency or toxicity symptoms were reported in creeping red fescue, Kentucky bluegrass, and perennial ryegrass having respective tissue Zn levels of 31.5, 22.3, and 30.6 ppm (84). A tissue Zn sufficiency range of 20–55 ppm was proposed by Jones (49). Others (80) suggested a minimal acceptable tissue level of 15 ppm of Zn. No deleterious effects from Zn tissue levels of 3,000 ppm were reported for bentgrass as a result of Zn applications (15).

Manganese

Wide ranges of tissue Mn have been reported in the literature. See Table 20.2 for examples, and note that levels reported by two researchers differed almost 10-fold. Tissue Mn in turfgrasses was reported to be highly dependent on soil pH (84), being much lower when soils were limed. A sufficiency range for tissue Mn of 25–150 ppm for turfgrasses has been proposed (49), with a minimal acceptable tissue level of 25 ppm (80).

Copper

Tissue Cu levels in turfgrasses tend to be substantially lower than Fe and Mn levels (Table 20.2). Tissue Cu levels of 8.0, 8.4, and 7.3 ppm were reported in perennial ryegrass, creeping red fescue, and Kentucky bluegrass, respectively, without any visual or measurable deficiencies (84). A range of 11–25 ppm of Cu has been reported in Kentucky bluegrass tissue during the growing season (39). A turfgrass sufficiency range of 5–20 ppm has been proposed (49), with a minimal acceptable tissue level of 5 ppm of Cu (80).

Boron

Relative to all other nutrients except Mo, B is required in very small amounts. Tissue B levels are reported in Table 20.2. A range of 9–27 ppm B has been reported in Kentucky bluegrass tissue during the growing season (39). Although Jones (49) suggested a turfgrass sufficiency range for tissue B of 10–60 ppm, no deficiencies or toxicities were noted by Turner (86) for perennial ryegrass, creeping red fescue, or Kentucky bluegrass having tissue B levels of 9.4, 9.2, and 7.9 ppm, respectively. A minimal acceptable tissue level of 3 ppm of B has been suggested (80).

Molybdenum

Turfgrasses require less Mo than any other essential nutrient. Few measurements have been made, however, to quantify deficient and toxic levels. Jones (49) stated that not enough is known regarding turfgrass Mo levels to determine a sufficiency range. A minimal acceptable tissue level of 2 ppm of Mo has been suggested (80). Tissue levels reported in Table 20.2 are close to or above

Table 20.2. Nutrient values for various turfgrass species and cultivars[a]

Species or cultivar	P (%) A	K (%) A	Ca (%) A	Ca (%) B	Mg (%) A	Mg (%) B	Fe (ppm)[b] A	Fe (ppm)[b] B	Zn (ppm) A	Zn (ppm) B	Mn (ppm) A	Mn (ppm) B	Cu (ppm) A	Cu (ppm) B	B (ppm) A	B (ppm) B	Mo (ppm) B
Colonial bentgrass	0.53	4.5	0.66	0.36	0.21	0.25	204	179	70	50	414	83	31	19	26	6	2.2
Creeping red fescue	0.54	3.6	0.49	0.39	0.17	0.24	111	266	54	30	185	54	25	20	26	7	2.7
Kentucky bluegrass	0.56	3.7	0.56	0.27	0.16	0.16	107	102	52	19	154	18	25	36	16	6	1.8
Annual bluegrass	0.57	3.6	0.64	...	0.20	...	135	...	78	...	250	...	26	...	36	...	...
Tall fescue	0.64	4.3	0.50	0.49	0.30	0.35	127	189	50	45	434	48	23	26	22	9	3.4
Perennial ryegrass	0.71	4.4	0.60	0.51	0.19	0.32	162	354	77	47	304	71	24	34	24	9	4.0
Creeping bentgrass	0.76	3.9	0.62	...	0.22	...	170	...	61	...	399	...	35	...	30	...	...
Marion Kentucky bluegrass	...	...	...	0.36	...	0.32	...	934	...	52	...	73	...	38	...	14	8.4
Meyer zoysiagrass	...	...	...	0.29	...	0.13	...	203	...	35	...	29	...	18	...	6	1.8
Bermudagrass	...	...	...	0.38	...	0.25	...	1,066	...	34	...	57	...	43	...	10	8.2

[a] Data in column A are from Waddington and Zimmerman (88) and in column B from Butler and Hodges (9).
[b] Parts per million values (mg/kg).

2.0. A range of 1.2–2.8 ppm of Mo in Kentucky bluegrass tissue during the growing season has been reported (39).

Selected References

1. Adams, W. A. 1980. Effects of nitrogen fertilization and cutting height on the shoot growth, nutrient removal and turfgrass composition of an initially perennial ryegrass dominant sports turf. Pages 343-350 in: Proc. Int. Turfgrass Res. Conf. 3rd. J. B. Beard, ed. International Turfgrass Society and American Society of Agronomy, Crop Science Society of America, and Soil Science Society of America, Madison, WI.

2. Adams, W. E., and Twersky, M. 1960. Effect of soil fertility on winter killing of coastal bermudagrass. Agron. J. 52:325-326.

3. Alexander, P. M., and Gilbert, W. B. 1963. Winter damage to bermuda greens. Golf Course Rep. 31(9):50-53.

4. Beard, J. B., and Rieke, P. E. 1966. The influence of nitrogen, potassium, and cutting height on the low temperature survival of grasses. In: Agronomy Abstracts. American Society of Agronomy, Madison, WI.

5. Bell, R. S., and DeFrance, J. A. 1944. Influence of fertilizers on the accumulation of roots from closely clipped bentgrasses and on the quality of the turf. Soil Sci. 58:17-24.

6. Bloom, J. R., and Couch, H. B. 1960. Influence of environment on diseases of turfgrasses. I. Effect of nutrition, pH, and soil moisture on Rhizoctonia brown patch. Phytopathology 50:532-535.

7. Brauen, S. E., Goss, R. L., Gould, C. J., and Orton, S. P. 1975. The effects of sulphur in combinations with nitrogen, phosphorus, and potassium on color and Fusarium patch disease of *Agrostis* putting green turf. J. Sports Turf Res. Inst. 51:83-91.

8. Brown, J. C. 1982. Summary of symposium on iron nutrition and interactions in plants. J. Plant Nutr. 5:987-1001.

9. Butler, J. D., and Hodges, T. K. 1967. Mineral composition of turfgrasses. HortScience 2:62-63.

10. Cahill, J. V., Murray, J. J., O'Neill, N. R., and Dernoeden, P. H. 1983. Interrelationships between fertility and red thread fungal disease of turfgrass. Plant Dis. 67:1080-1083.

11. Canaway, P. M. 1984. The response of *Lolium perenne* (perennial ryegrass) turf grown on sand and soil to fertilizer nitrogen. II. Above-ground biomass, tiller numbers and root biomass. J. Sports Turf Res. Inst. 60:19-26.

12. Carroll, J. C., and Welton, F. A. 1939. Effects of heavy and late applications of nitrogenous fertilizers on the cold resistance of Kentucky bluegrass. Plant Physiol. 14:297-308.

13. Carrow, R. N., Johnson, B. J., and Landry, Jr., G. W. 1988. Centipedegrass response to foliar application of iron and nitrogen. Agron. J. 80:746-750.

14. Cheesman, J. H., Roberts, E. C., and Tiffany, L. H. 1965. Effects of nitrogen level and osmotic pressure of the solution on incidence of *Puccinia graminis* and *Helminthosporium sativum* infection in 'Merion' Kentucky bluegrass. Agron. J. 57:599-602.

15. Christians, N. E. 1984. Eagle iron studies. 1984 Iowa Turfgrass Res. Rep., Iowa State University, Ames.

16. Christians, N. E., Diesburg, K. L., and Nus, J. L. 1985.

17. Christians, N. E., Martin, D. P., and Karnok, K. J. 1981a. The interrelationship among nutrient elements applied to calcareous sand greens. Agron. J. 73:929-933.

18. Christians, N. E., Martin, D. P., and Karnok, K. J. 1981b. The interaction among nitrogen, phosphorus, and potassium on the establishment, quality and growth of Kentucky bluegrass (*Poa pratensis* L. 'Merion'). Pages 341-348 in: Proc. Int. Turfgrass Res. Conf. 4th. R. W. Sheard, ed. International Turfgrass Society and Ontario Agricultural College, University of Guelph, Guelph, Ontario, Canada.

19. Christians, N. E., Martin, D. P., and Wilkinson, J. F. 1979. Nitrogen, phosphorus, and potassium effects on quality and growth of Kentucky bluegrass and creeping bentgrass. Agron. J. 71:564-567.

20. Cook, R. N., Engel, R. E., and Bachelder, S. 1964. A study of the effect of nitrogen carriers on turfgrass disease. Plant Dis. Rep. 48:254-255.

21. Couch, H. B. 1973. Diseases of Turfgrasses. 2nd ed. Robert E. Kreiger, Huntington, NY.

22. Danneberger, T. K., Vargas, J. M., Rieke, P. E., and Street, J. R. 1983. Anthracnose development on annual bluegrass in response to nitrogen carriers and fungicide application. Agron. J. 75:35-38.

23. Davidson, R. M., Jr., and Goss, R. L. 1972. Effects of P, S, N, lime, chlordane, and fungicides on ophiobolus patch disease on turf. Plant Dis. Rep. 56:565-567.

24. Deal, E. D., and Engel, R. E. 1965. Iron, manganese, boron, and zinc: Effects on growth of Merion Kentucky bluegrass. Agron. J. 57:533-555.

25. Dernoeden, P. H. 1987. Management of take-all patch of creeping bentgrass with nitrogen sulfur, and phenyl mercury acetate. Plant Dis. 71:226-229.

26. Dernoeden, P. H., and Jackson, N. 1980. Managing yellow tuft disease. J. Sports Turf Res. Inst. 56: 9-17.

27. Dest, W. M., and Allinson, D. W. 1981. Influence of nitrogen, and phosphorus fertilization on the growth and development of *Poa annua* L. (annual bluegrass). Pages 325-332 in: Proc. Int. Turfgrass Res. Conf. 4th. R. W. Sheard, ed. International Turfgrass Society and Ontario Agricultural College, University of Guelph, Guelph, Ontario, Canada.

28. Dudeck, A. E., Peacock, C. H., and Freeman, T. E. 1985. Response of selected bermudagrasses to nitrogen fertilization. Pages 495-504 in: Proc. Int. Turfgrass Res. Conf. 5th. F. Lemaire, ed. Institut National de la Recherche Agronomique, Paris.

29. Endo, R. M. 1966. Control of dollar spot of turfgrass by nitrogen and its probable bases. (Abstr.) Phytopathology 56:877.

30. Escritt, J. R., and Legg, D. C. 1970. Fertilizer trials at Bingley. Pages 185-190 in: Proc. Int. Turfgrass Res. Conf. 1st. Sports Turf Research Institute, Bingley, England.

31. Fry, J. O., and Dernoeden, P. H. 1987. Growth of zoysiagrass from vegetative plugs in response to fertilizers. J. Am. Soc. Hortic. Sci. 112:286-289.

32. Gilbeault, V. A., and Hanson, D. 1980. Perennial ryegrass

Effects of nitrogen fertilizer and fall topdress on the spring recovery of *Agrostis palustris* Huds. ('Penncross' creeping bentgrass) greens. Pages 459-467 in: Proc. Int. Turfgrass Res. Conf. 5th. F. Lemaire, ed. Institut National de la Recherche Agronomique, Paris.

mowing quality and appearance in response to three nitrogen regimes. Pages 39-43 in: Proc. Int. Turfgrass Res. Conf. 3rd. J. B. Beard, ed. International Turfgrass Society and American Society of Agronomy, Crop Science Society of America, and Soil Science Society of America, Madison WI.

33. Gilbert, W. B., and Davis, D. L. 1971. Influence of fertility ratios on winter-hardiness of bermudagrass. Agron. J. 63:591-593.

34. Goss, R. L. 1969. Some inter-relationships between nutrition and turf disease. Pages 351-361 in: Proc. Int. Turfgrass Res. Conf. 1st. Sports Turf Research Institute, Bingley, Yorkshire.

35. Goss, R. L., Brauen, S. E., and Orton, S. P. 1975. The effects of N, P, K, and S on *Poa annua* L. in bentgrass putting green turf. J. Sports Turf Res. Inst. 51:74-82.

36. Goss, R. L., Brauen, S. E., and Orton, S. P. 1979. Uptake of sulfur by bentgrass putting green turf. Agron. J. 71:909-913.

37. Goss, R. L., and Gould, C. J. 1968. Some interrelationships between fertility levels and Fusarium patch disease in turfgrasses. J. Sports Turf Res. Inst. 44:19-26.

38. Gould, C. J., Miller, V. L., and Goss, R. L. 1967. Fungicidal control of red thread disease of turfgrass in western Washington. Plant Dis. Rep. 51:215-219.

39. Hall, J. R., and Miller, R. W. 1974. Effect of phosphorus, season and method of sampling on foliar analysis of Kentucky bluegrass. Pages 155-171 in: 1973 Proc. Int. Turfgrass Res. Conf. 2nd. E. C. Roberts, ed. American Society of Agronomy and Crop Science Society of America, Madison, WI.

40. Harivandi, M. A., and Butler, J. D. 1980. Iron chlorosis of Kentucky bluegrass cultivars. HortScience 15:496-497.

41. Hawes, D. T., and Decker, A. M. 1977. Healing potential of creeping bentgrass as affected by nitrogen and soil temperature. Agron. J. 69:212-214.

42. Holt, C. C., and Davis, R. L. 1948. Differential responses of Arlington and Norbeck bentgrasses to kinds and rate of fertilizers. Agron. J. 40:282-284.

43. Horst, G. L. 1984. Iron nutrition for warm season grasses. Ground Maintenance July:44.

44. Horst, G. L., Baltensperger, A. A., and Firkner, M. D. 1985. Effects of N and growing season on root-rhizome characteristics of turf-type bermudagrasses. Agron. J. 77:237-242.

45. Hull, R. J., Jackson, N., and Skogley, C. R. 1979. Influence of nitrogen on stripe smut severity in Kentucky bluegrass turf. Agron. J. 71:553-555.

46. Hylton, L. O., Jr., Ulrich, A., Cornelius, D. R., and Okhi, K. 1965. Phosphorus nutrition of Italian ryegrass relative to growth, moisture content, and mineral constituents. Agron. J. 57:505-508.

47. Johnson, B. J. 1984. Influence of nitrogen on recovery of bermudagrass (*Cynodon dactylon*) treated with herbicides. Weed Sci. 32:819-823.

48. Johnson, B. J., and Bowyer, T. H. 1982. Management of herbicide and fertility levels on weeds and Kentucky bluegrass turf. Agron. J. 74:845-850.

49. Jones, J. R., Jr. 1980. Turf analysis. Golf Course Manage. 48(1):29-32.

50. Juska, F. V. 1959. Response of Meyer zoysia to lime and fertilizer treatments. Agron. J. 51:81-83.

51. Juska, F. V., and Hanson, A. A. 1966. Nutritional requirements of *Poa annua* L. Page 35 in: Agronomy Abstracts. American Society of Agronomy, Madison, WI.

52. Juska, F. V., Hanson, A. A., and Erickson, C. J. 1965. Effects of phosphorus and other treatments on the development of red fescue, Merion, and common Kentucky bluegrass. Agron. J. 57:75-78.

53. Juska, F. V., and Murray, J. M. 1974. Performance of bermudagrass in the transition zone as affected by potassium and nitrogen. Pages 149-154 in: 1973 Proc. Int. Turfgrass Res. Conf. 2nd. E. C. Roberts, ed. American Society of Agronomy and Crop Science Society of America, Madison, WI.

54. Juska, F. V., Tyson, J., and Harrison, C. M. 1955. The competitive relationships of Merion Kentucky bluegrass as influenced by various mixtures, cutting heights and levels of nitrogen. Agron. J. 47:513-518.

55. Kamon, Y. 1974. Magnesium deficiency in zoysiagrass. Pages 145-148 in: 1973 Proc. Int. Turfgrass Res. Conf. 2nd. E. C. Roberts, ed. American Society of Agronomy and Crop Science Society of America, Madison, WI.

56. Keisling, T. C. 1980. Bermudagrass rhizome initiation and longevity under differing potassium nutritional levels. Commun. Soil Sci. Plant Anal. 11:629-635.

57. Kohlmeier, G. P., and Eggens, J. L. 1983. The influence of wear and nitrogen on creeping bentgrass growth. Can. J. Plant Sci. 63:189-193.

58. Kurtz, K. W. 1981. Use of 59 Fe in nutrient solution cultures for selecting and differentiating Fe-efficient and Fe-inefficient genotypes in zoysiagrass. Pages 267-275 in: Proc. Int. Turfgrass Res. Conf. 4th. R. W. Sheard, ed. International Turfgrass Society and Ontario Agricultural College, University of Guelph, Guelph, Ontario, Canada.

59. Leyer, C., and Skirde, W. 1980. Effects of nitrogen fertilizer levels and wear tolerance of sports turf. Z. Veget. Landsch. Sports 3:25-31.

60. Love, J. R. 1962. Mineral deficiency symptoms on turfgrass. I. Major and secondary nutrient elements. Wis. Acad. Sci. Arts Lett. 51:135-140.

61. Madsen, J. P., and Hodges, C. F. 1980. Nitrogen effects on the pathogenicity of *Drechslera sorokiniana* and *Curvularia geniculata* on germinating seed of *Festuca rubra*. Phytopathology 70:1033-1036.

62. Markland, F. E., Roberts, E. C., and Frederick, L. R. 1969. Influence on nitrogen fertilizers on Washington creeping bentgrass *Agrostis palustris* Huds. II. Incidence of dollar spot (*Sclerotinia homecarpa*) infection. Agron. J. 61:701-705.

63. McCaslin, B. D., Samson, R. F., and Baltensperger, A. A. 1981. Selection for turf-type bermudagrass genotypes with reduced iron chlorosis. Commun. Soil Sci. Plant Anal. 12:189-204.

64. McVey, G. R. 1967. Response of seedlings to various phosphorus sources. Page 53 in: Agronomy Abstracts. American Society of Agronomy, Madison. WI.

65. Mehall, B. J., Hull, R. J., and Skogley, C. R. 1983. Cultivar variation in Kentucky bluegrass: P and K nutritional factors. Agron. J. 75:767-772.

66. Menn, W. G., and McBee, G. G. 1970. A study of certain nutritional requirements for Tifgreen bermudagrass (*Cynodon dactylon* × *C. transvaalensis*) utilizing a hydroponic system. Agron. J. 62:192-194.

67. Minner, D. D., and Butler, J. D. 1986. Iron nutrition of turfgrass. Pages 125-148 in: Advances in turfgrass fertility. B. G. Joyner, ed. Chemlawn Services Corporation, Columbus, OH.

68. Monroe, C. A., Coorts, G. D., and Skogley, C. R. 1969. Effects of nitrogen-potassium levels on the growth and chemical composition of Kentucky bluegrass. Agron. J. 61:294-296.

69. Moore, L. D., Couch, H. B., and Bloom, J. R. 1961. Influence of nutrition, pH, soil temperature, and soil moisture on Pythium blight of Highland bentgrass. (Abstr.) Phytopathology 51:578.

70. Murray, J. J., Klingman, D. L., Nash, R. G., and Woolson, E. A. 1983. Eight years of herbicide and nitrogen fertilizer treatments on Kentucky bluegrass (*Poa pratensis*) turf. Weed Sci. 31:825-831.

71. Nittler, L. W., and Kenny, W. J. 1971. Cultivar differences among calcium deficient Kentucky bluegrass seedlings. Agron. J. 64:73-75.

72. Oertli, J. R., Lunt, O. R., and Youngner, V. B. 1961. Boron toxicity in several turfgrass species. Agron. J. 53:262-265.

73. Palmertree, H. D., Ward, C. Y., and Pluenneke, R. H. 1974. Influence of mineral nutrition on the cold tolerance and soluble protein fraction of centipedegrass. Pages 500-507 in: 1973 Proc. Int. Turfgrass Res. Conf. 2nd. E. C. Roberts, ed. American Society of Agronomy and Crop Science Society of America, Madison, WI.

74. Pellett, R. M., and Roberts, E. C. 1963. Effects of mineral nutrition on high temperature induced growth retardation of Kentucky bluegrass. Agron. J. 55:474-476.

75. Pritchett, W. L., and Horn, G. C. 1966. Fertilization fights turf disorders. Better Crops Plant Food 50(3):22-25.

76. Ryan, J., Stroehlein, J. L., and Miyamoto, S. 1975. Sulfuric acid applications to calcareous soils: Effects on growth and chlorophyll content of common bermudagrass in the greenhouse. Agron. J. 67:633-637.

77. Sartain, J. B., and Dudeck, A. E. 1980. Yield and nutrient accumulation of Tifway bermudagrass and overseeded ryegrass as influenced by applied nutrients. Agron. J. 74:488-491.

78. Schmidt, R. E., and Blaser, R. E. 1967. Effect of temperature, light, and nitrogen on growth and metabolism of 'Cohansey' bentgrass (*Agrostis palustris* Huds). Crop Sci. 7:447-451.

79. Schmidt, R. E., and Brueninger, J. M. 1981. The effects of fertilization on recovery of Kentucky bluegrass turf from summer drought. Pages 333-340 in: Proc. Int. Turfgrass Res. Conf. 4th. R. W. Sheard, ed. International Turfgrass Society and Ontario Agricultural College, University of Guelph, Guelph, Ontario, Canada.

80. Snyder, G. H., and Augustin, B. J. 1986. Managing micronutrient applications on Florida turfgrasses. Pages 149-179 in: Advances in turfgrass fertility. B. G. Joyner, ed.

Chemlawn Services, Columbus, OH.

81. Snyder, V., and Schmidt, R. E. 1974. Nitrogen and iron fertilization of bentgrass. Pages 176-185 in: 1973 Proc. Int. Turfgrass Res. Conf. 2nd. E. C. Roberts, ed. American Society of Agronomy and Crop Science Society of America, Madison, WI.

82. Sturkie, D. G., and Rouse, R. D. 1967. Response of zoysia and Tiflawn bermuda to P and K. Page 54 in: Agronomy Abstracts. American Society of Agronomy, Madison, WI.

83. Turgeon, Stone, G. G., and Peck, T. R. 1979. Crude protein levels in turfgrass clippings. Agron. J. 71:229-232.

84. Turner, T. R. 1980. Soil test calibration studies for turfgrasses. Ph.D. diss. The Pennsylvania State University, University Park, PA. Diss. Abstr. 80-24499.

85. Turner, T. R., and Waddington, D. V. 1983. Soil test calibration for establishment of turfgrass monostands. Soil Sci. Soc. Am. J. 47:1161-1166.

86. Turner, T. R., Waddington, D. V., and Watschke, T. L. 1979. The effect of fertility levels on dandelion and crabgrass encroachment of Merion Kentucky bluegrass. Proc. Annu. Meet. Northeast. Weed Sci. Soc. 33:280-286.

87. Waddington, D. V., Turner, T. R., Duich, J. M., and Moberg, E. L. 1978. Effect of fertilization on 'Penncross' creeping bentgrass. Agron. J. 70:713-718.

88. Waddington, D. V., and Zimmerman, T. L. 1972. Growth and chemical composition of eight grasses grown under high water table conditions. Commun. Soil Sci. Plant Anal. 3:329-337.

89. Watschke, T. L., and Waddington, D. V. 1974. Effect of nitrogen source, rate, and timing on growth and carbohydrates of 'Merion' Kentucky bluegrass. Agron. J. 66:691-696.

90. Watschke, T. L., and Waddington, D. V. 1975. Effect of nitrogen fertilization on the recovery of Merion' Kentucky bluegrass from scalping and wilting. Agron. J. 67:559-563.

91. Westfall, R. T., and Simmons, J. A. 1971. Germination and seeding development of Windsor Kentucky bluegrass as influenced by phosphorus and other nutrients. Page 52 in: Agronomy Abstracts. American Society of Agronomy, Madison, WI.

92. Wood, J. R., and Duble, R. L. 1976. Effects of nitrogen and phosphorus on establishment and maintenance of St. Augustinegrass. Texas Agric. Exp. Stn. PR-3368C.

93. Woolhouse, A. R. 1986. The assessment of perennial ryegrass cultivars for susceptibility to red thread disease. J. Sports Turfgrass Res. Inst. 62:147-152.

94. Wu, L., Huff, D. R., and Johnson, J. M. 1981. Metal tolerance of bermudagrass cultivars. Pages 35-40 in: Proc. Int. Turfgrass Res. Conf. 4th. R. W. Sheard, ed. International Turfgrass Society and Ontario Agricultural College, University of Guelph, Guelph, Ontario, Canada.

95. Yust, A. K., Wehner, D. J., and Fermanian, T. W. 1984. Foliar application of N and Fe to Kentucky bluegrass. Agron. J. 76:934-938.

Related Publications

Additional publications on plant nutrition and plant symptoms may be of value.

The book listed below by Chapman was an extremely thorough and early compilation of critical nutrient levels. Published in 1966, the information presented is still of value. The book by Jones is a recent and very excellent compilation of interpretative plant analysis data on 306 individual crops. Any person using plant analysis as a diagnostic tool needs this publication.

Diagnostic Criteria for Plants and Soil, by Homer D. Chapman. (University of California, Riverside, 1966.)

Plant Analysis Handbook, by J. Benton Jones, Jr., B. Wolf, and H. A. Mills. (Micro-Macro Publishing, Athens, GA, 1991.)

An understanding of plant nutrition is helpful in the diagnosis of plant problems. The following are five excellent publications.

Detecting Mineral Nutrient Deficiencies in Tropical and Temperate Crops, by D. L. Plucknett and H. B. Sprague. (Westview Tropical Agriculture Series, No. 7. Westview Press, Boulder, CO, 1989.)

Diagnosis of Mineral Disorders in Plants. Vol. 1, Principles, by C. Bould, E. H. Hewitt, and P. Needham. (Chemical Publishing, New York, 1984.)

Diagnosis of Mineral Disorders in Plants. Vol. 2, Vegetables, by A. Scarfe and Mary Turner. (Chemical Publishing, New York, 1984.)

Mineral Nutrition of Higher Plants, by Horst Marschner. (Academic Press, London, 1986.)

Principles of Plant Nutrition, by K. Mengel and E. A. Kirkby. (International Potash Institute, Berne, Switzerland, 1978.)

It would not be proper to fail to mention the "bible" on nutrient deficiency symptoms, which is now out of print.

Hunger Signs in Crops: A Symposium, edited by Howard B. Sprague. (David McKay, New York, 1964.)

Index